THE MAKING OF A CONTINENT

THE MAKING OF

COLOR ILLUSTRATIONS BY   GARY   HINCKS

DESIGN BY BETTY BINNS

A CONTINENT

TEXT AND PHOTOGRAPHS BY RON REDFERN

BBC BOOKS

"What is the use of a book," thought Alice,
"without pictures or conversations?"
LEWIS CARROLL

This book is dedicated to the men and women
of science who told me where and when and why.
I shall forever be in their debt.

Published by the British Broadcasting Corporation
35 Marylebone High Street, London W1M 4AA
First Published 1983

Published simultaneously in Canada by
Fitzhenry & Whiteside, Ltd., Toronto

Printed in The Netherlands by Royal Smeets
Offset B. V., Weert

ISBN 0-563-20235-1

83 84 85 86 87 5 4 3 2 1

CONTENTS

INTRODUCTION

ALTHOUGH logic and fact are together considered the mainstay of scientific progress, some apparently illogical ideas have resulted in quantum leaps in human knowledge. For instance, in the 16th century Nicolaus Copernicus postulated the herctical idea of a solar system that ultimately revolutionized the science of astronomy. In the 19th century Charles Darwin's then unacceptable concept of evolutionary processes eventually revolutionized biological science. And also in the 19th century Louis Agassiz promoted the preposterous idea that past ages of ice had shaped northern landscapes: an idea that fathered the sciences of glaciology and meteorology. Early in the present century Alfred Wegener contended that continents move about on the Earth's surface and had once formed a single continent, which he named "Pangea." Wegener sparked a divisive controversy that led to the recent establishment of a new geological concept: "plate tectonics." Collectively these hypotheses have completely changed the course of both science and human thought. Perhaps one should conclude from such examples that the quality of any good hypothesis can be measured by its degree of outrageousness.

This book is the result of a personal quest. I wanted better to understand the implications of the new tectonic science and how this concept of the Earth's crustal genesis has brought about a revolution not just in geology but also in the interpretation of the facts of natural science as a whole. To gain a perspective, I concentrated on one particular part of the lithosphere, the North American Plate, one of a number that cover Earth's surface. By understanding the North American continent's geological history, I hoped to discover for myself to what degree this history had determined the course of North American life—hence the title of the book: *The Making of a Continent.*

At the end of the project my conclusion is inescapable: Evolution of life upon the tectonically isolated North American continent has indeed been profoundly affected by its geological history. In fact, this continent appears to me to exemplify the general effect of Earth dynamics upon life's evolution on the upper crust. During the process of reaching this conclusion I acquired at least a feel for the tectogenesis of other complex structures — the vast Eurasian continent, for example. The consequence is that I now find myself looking at all landscapes and seascapes with entirely different eyes and at a physical map of the world with renewed interest. Most surprising to me has been the discovery of the extent to which the tectogenetics of North America and Europe are inextricably linked.

The circumpolar view of Earth that appears with this "Introduction" is bereft of its oceans and ice caps and is set at a carefully considered angle. In fact nearly all Gary Hincks's illustra-

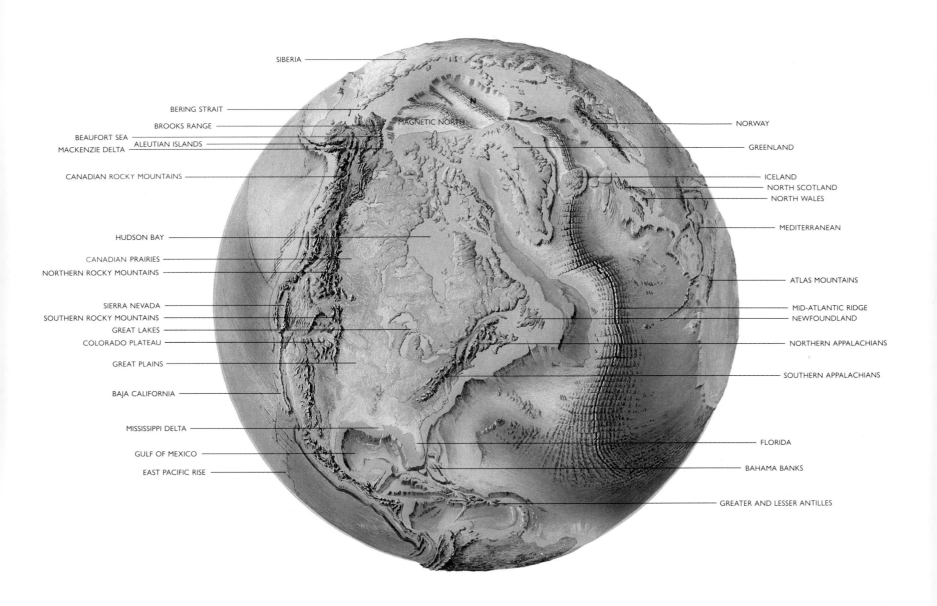

SIBERIA

BERING STRAIT

BROOKS RANGE

BEAUFORT SEA
ALEUTIAN ISLANDS
MACKENZIE DELTA

CANADIAN ROCKY MOUNTAINS

HUDSON BAY

CANADIAN PRAIRIES

NORTHERN ROCKY MOUNTAINS

SIERRA NEVADA

SOUTHERN ROCKY MOUNTAINS

GREAT LAKES

COLORADO PLATEAU

GREAT PLAINS

BAJA CALIFORNIA

MISSISSIPPI DELTA

GULF OF MEXICO

EAST PACIFIC RISE

MAGNETIC NORTH

NORWAY

GREENLAND

ICELAND
NORTH SCOTLAND
NORTH WALES

MEDITERRANEAN

ATLAS MOUNTAINS

MID-ATLANTIC RIDGE
NEWFOUNDLAND

NORTHERN APPALACHIANS

SOUTHERN APPALACHIANS

FLORIDA

BAHAMA BANKS

GREATER AND LESSER ANTILLES

tions in the book are linked to this key illustration in the sense that, no matter to what degree of detail, they are drawn from the same circumpolar view. As the story develops, some graphics show the globe in partial cross section to accent the planet's geophysical relationship to the structure of the North American continent. Other cross sections focus on certain features of the continent that have a particularly strong geological significance. And throughout the book smaller color diagrams illustrate a variety of tectonic mechanisms that in one way or another contribute to change.

There is a two-fold objective in this scheme: The first is to project a view of the physical globe that emphasizes the relationship of the North American plate to the Eurasian and African plates, with the Mid-Atlantic Ridge separating them. The second objective is to enable a reader perhaps unfamiliar with this topography to relate the subject of any chapter or of any detailed Hincks illustration to the whole circumpolar view.

Continents have common factors: Each has a nucleus, a stable platform, an unstable belt, a continental shelf, and other anatomical parts. I have used some of these rather exact but very descriptive geological terms for chapter titles. Continents can also be divided into physiographical provinces, each with its distinctive geological history. In North America there are forty or more such provinces and sub-provinces. In each chapter I have taken one province representative of a particular aspect of continental anatomy and have provided a synthesis of its geological history — so far as it is known — and its relationship to life's development in that particular province. Depending on the locality, I have sometimes emphasized natural history, sometimes aboriginal history, and always the province's modern history — by which I mean non-political human history in terms of the last few centuries.

Each chapter takes the reader through a sequence of events from which I have tried to construct a synthesis of a continental geophysical machine at work, as well as a vocabulary to describe the process. The order of subject presentation therefore requires a physiographic progression and *not* the more familiar geographic progression of a travel-

ogue (to which role this book has no pretense). I move from one part of continental anatomy to the next logical part, and this sometimes requires a quite considerable geographical step: Each chapter therefore opens with a miniature circumpolar view with the physiographic area discussed in the chapter emphasized. The first chapter discusses the probable reasons and the proofs of continental movement. I then progress to the nucleus of a continent and its stable parts and its not-so-stable parts outward to the freshly made edges and later to the worn parts. Having created an image of continental anatomy, I deal with a continent's physiology: how it drains, how it forms new rock, and what traumas affected its past and threaten its present and future. It follows that chapters cannot be read out of sequence, for I depend on the reader's growing familiarity with a new language, a new science, and a developing story.

We humans are blessed with a large visual cortex and a unique capacity for constructive thought: We therefore respond to ideas that are illustrated with easier understanding and greater enjoyment than to the written word alone. The book's color photographs, color illustrations, line drawings, and maps are integral parts of my synthesis. They do not simply augment the textual treatment — they are inseparable from it. But the photography serves an extra purpose. It reflects my personal view of a continental landscape. It is an interpretative view — whatever the subject, I first had to understand its nature and its context in the overall story before I could plan the shot. But of course I also tried to take photographs that would excite the imagination, and this was a keenly felt responsibility. Few people have the opportunity of visiting the whole extent of a continent and then of presenting their pictorial view of it in one volume.

The language of any science is strange to many laymen but it is a necessary shorthand for the scientist. In a book of this character one cannot repeatedly coin phrases and sentences to describe the same perhaps complicated sequence of events when specialists have invented one perfectly good word or phrase to save the need. I have limited such technical words to an essential dozen or so, and in every case have interpreted its meaning in preliminary phrases. The necessary vocabulary has been built gradually, as the logic of the story requires. Even so, the index has been prepared so that the page number of the first definitive

use of more technical words appears in boldface for easy reference.

But perhaps the most confusing geological language to the uninitiated relates the division of geologic time. If the Latin names for the eons, eras, periods, and their subdivisions are familar friends, you will find them used in conjunction with the drawings: the time-spirals, the paleogeographic diagrams, and the charts. But for those not so familiar with the expressions — and remembering well my own confusion at first aquaintance with them — I hope to have reduced the problem by using three straightforward terms in their place: simply, "pre-Pangean," "Pangean," and "post-Pangean." The one additional word I have used throughout is "Precambrian," the name used to describe pre-Pangean time before the first invertebrate life appeared in the fossil record.

During the course of the two years' preparatory work for this book several hundred scientists and specialists gave generously of their time to explain, to help, and to encourage. Although individual acknowledgment precedes the bibliography, in this "Introduction" I would like to pay a special tribute to the people who gave generously of their time. Most are leaders in widely different aspects of natural science, but they have one important attribute in common: All are field-scientists who know their subject from direct experience of it. They therefore contributed to both the technical and the photographic aspects of this book: They told me where and when and why. They are most kind and patient people; without their direct and willing help this book would have been impossible to complete.

The science of plate tectonics is space-age geology and marvelously interesting as well as profoundly important to our understanding of a complex parent organism — our dynamic Earth. It is equally as fascinating a place as Mars or Saturn or Jupiter, but it doesn't get quite so much publicity. Unlike those other planets we see the Earth's surface every day of our lives and rather take it for granted. We shouldn't — as I very much hope this modest entertainment demonstrates.

Ron Redfern
April, 1983

CONTINENTS APART

ONE glorious mid–October morning I sat contemplating the view from a large block of limestone perched precipitously on the South Rim of the Grand Canyon of the Colorado in Arizona. Behind me, in the chilly shadow of the pinyon pines of Shoshone Point, I could hear an occasional crackle as logs burned for a pleasant alfresco meal. There was no other sound. No wind in the trees. Not a bird stirred. Not a living thing seemed to move under the cloudless sky. Just crystalline air and one of the world's wonders spread peacefully below. A moment of "bliss unmixed with earthly lot."

As I gazed into the depths of the canyon, I recalled the sensation of weaving a way by helicopter between the crested buttes, tracing the path of Upper Granite Gorge and the Colorado River, searching for a vantage point from which to photograph a panorama of the Unkar section of the canyon. The sedimentary rocks of Unkar are very old, about 800 million years old, and some contain fossils of the oldest life forms known in the canyon. Even so these rocks are still 1,000 million years younger than those of Granite Gorge lower down the river — almost 2,000 million years in the past. I reflected soberly on the brevity of my allotted span.

Some strange association of ideas caused me to remember a boyhood hero of mine, the rambunctious Professor George Edward Challenger in Arthur Conan Doyle's *The Lost World*: Challenger and his companions had climbed to the top of some imaginary cliffs to an island in the sky, where fearsome reptiles from the past had somehow survived extinction. The connection came to me. Beyond Unkar, stupendous tiered cliffs rise majestically toward the eastern horizon to form an even ridge between Cape Solitude and Desert View. Over the ridge and out of sight lies the Painted Desert, once a small part of a real lost world, a vast continent on which giants of a reptilian order roamed, only to become extinct as today's continents tore away from one another and slowly moved apart. For me the barren but vividly colorful badlands of the Painted Desert have always been a time window through which I could see the landscape of that lost world, the world of supercontinental Pangea.

But what, I wondered, is the anatomy of a continent? The ocean floors are lower than the continental masses by many thousands of meters. Why should this be so over seven-tenths of Earth's surface when it seems reasonable to suppose that rocks would conform to a general level? And how did Pangea itself form in the first place? Did other supercontinents precede it? And if so, what happened to them?

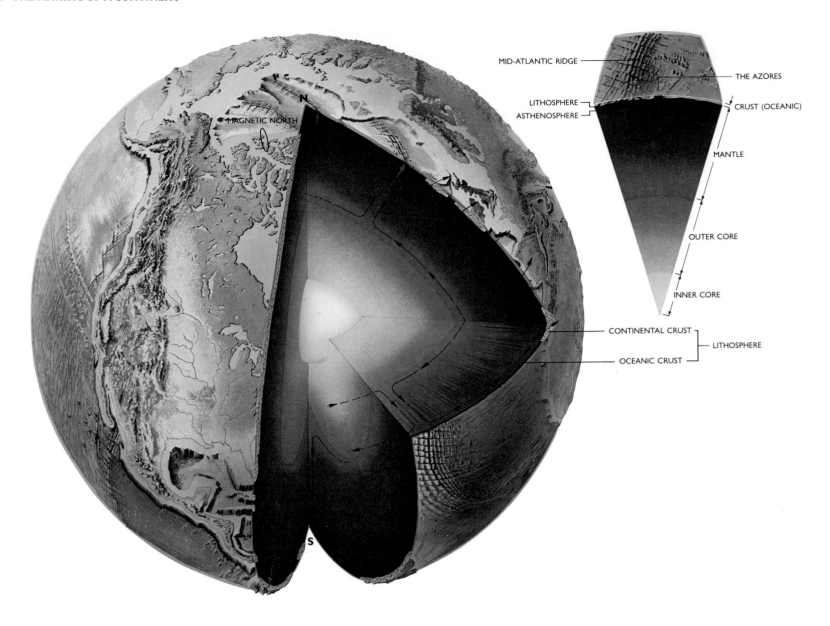

MID-ATLANTIC RIDGE

THE AZORES

LITHOSPHERE
ASTHENOSPHERE

CRUST (OCEANIC)

MANTLE

OUTER CORE

INNER CORE

MAGNETIC NORTH

N

S

CONTINENTAL CRUST
LITHOSPHERE
OCEANIC CRUST

That the continents on both sides of the Mid-Atlantic Ridge move away from each other can be proved in many ways, for instance, by the geological correlation between now opposing but once adjacent landmasses. But the mechanisms that cause movements of the continents and the ocean floors are not clearly understood and are at present the subject of intense research.

There are two main hypotheses to explain the general movement of the Earth's upper crust. The one illustrated here suggests that the very hot magma beneath the Earth's lithosphere circulates convectively, just as any liquid or semiliquid circulates as it is heated. Such convection can be visualized as a series of irregular-sized and perhaps deformed cells that interact with each other around the central core of Earth. The intricate interrelated movement of the cells acts upon the solid upper crust of the Earth, causing it too to move intricately. Because the Earth's surface is cold and solid (so the hypothesis goes), it is fractured into huge plates that separate, collide, scrape, and generally interact with each other, according to the movement of convective cells beneath.

The second hypothesis is sometimes referred to as the "tablecloth" model. The ocean-crust edges of some plates descend beneath an opposing plate. Ocean plates are lighter near their hot spreading ridges than they are at their distant cold (and therefore more dense) edges. The descending ocean plate gradually sinks into the regions below the upper crust, where it melts and is reabsorbed. The tablecloth model suggests that this subducting plate descends as a consequence of gravitational pull and not by the action of convection cells beneath it; the plate sinks as a consequence of its own weight at its dense edges — like a tablecloth slipping off a polished table top once most of the weight is over the table edge.

But most of all, how does a supercontinent begin to rift and how do the pieces move apart? What effect do such movements have on the shaping of continental landscapes, on hot climates and ice ages, on the evolution of life in general and humanity's relationship with the upper crust of Earth in particular?

The questions crowded in: Surely there must be other places that are like Grand Canyon and the Painted Desert, in the sense of being records of past events. I recalled Alfred Wegener, the scientist who had died a lone death from exposure on the Greenland ice cap in 1930 and was the author of *The Origin of Continents and Oceans*. He was the meteorologist whose hypothesis of continental drift, presented in Germany in 1912, had first startled and then divided the world of geological sciences for half a century. I had read that, if Wegener had concentrated his studies on Iceland instead of on Greenland, he might well have found the evidence he sought in support of his ideas. Perhaps Iceland would be the place for me to start? It is midway between the Faeroe Islands and eastern Greenland, the nearest extremities of the European and North American continents. If I could understand how Europe and North America had moved apart and then trace the development of North America as a separate continent, I should perhaps begin to accumulate a few answers. Such a venture would mean traveling from Arctic plains and permafrost to coral islands in tropical seas. It would mean visiting some of the largest human conurbations on Earth and some of the most remote uninhabited places. It was a daunting task to contemplate but an exciting prospect to speculate on. At that moment a call from the depths of the pinyon forest brought me back to reality. That lunch was ready.

Iceland is a volcanic island in the extreme north of the Atlantic Ocean; its northern shore is just short of the Arctic Circle, the imaginary line of latitude in the northern hemisphere that marks the summertime limit of perpetual day and the wintertime limit of perpetual night. When the Atlantic Ocean is graphically

drained away, Iceland can be clearly seen to interrupt the submarine continuity of a patterned ridge on the seabed. The ridge starts near the North Pole and separates Northern Europe from North America between the coastlines of Norway, the British Isles, and Greenland. At the intersection with an imaginary line joining the British Isles to Newfoundland the Mid-Atlantic Ridge, as it is called, jogs sharply eastward, but after this the ridge continues a generally southward trend halfway around the Earth, almost to the Antarctic, on a line roughly equidistant from the continents on either side. When it reaches the latitude of Cape Horn, the ridge joins another ocean ridge which runs at a right angle and continues into the Indian Ocean.

The ridge is a spreading zone, a region from which the seabed separates in opposite directions. In effect the Mid-Atlantic Ridge is a center line from which the extent of the divergence of continents on opposite sides of the ocean can be judged. The mechanism of seafloor spreading clearly is critical for understanding continental drift. Wegener had proposed that the continents ploughed their way through the ocean floor, and this one aspect of an already outrageous hypothesis caused the most vigorous opposition to his concepts. Iceland is one of the few places on Earth where a normally submarine spreading zone rises above the surface of the sea so that it provides a special opportunity for geologists to observe and study the very processes of separation that I wanted to understand better. This is why Iceland had to be the starting point in my adventure. I would talk to some of the geophysicists who spend much of their time studying the phenomenon of rifting in Iceland and then drive from south to north across the country along the ridge itself, if this were practical, to see it first hand. This would be the best way, I thought, to begin to understand how Pangea split into the present continents.

There is no apparent reason why there should be a volcanic island astride the Mid-Atlantic Ridge. There are many other deep-ocean volcanic systems that reach the surface of the sea, the Hawaiian Islands and the Azores being perhaps the two best known; but none of these island chains is coincident with a spreading ridge. In fact it is believed that such volcanic islands form as a conse-

quence of a stationary hot spot below the Earth's solid upper crust in the extremely hot semisolid region called the "mantle"; such hot spots are therefore called "mantle plumes." The mantle plume remains geographically fixed as the ocean bed moves over it. The interaction of plume and moving seabed results in the formation of a chain of "shield" volcanoes. The main volcanic activity is always on the trailing island, which is a huge erupting shield volcano with its peak above the sea and a line of extinct volcanoes ahead of it.

It is a mystery why there should be stationary mantle plumes below the Earth's solid upper regions, the lithosphere, burning holes through to the surface. The theories put forward to explain such plumes are considered to be highly speculative. One such theory suggests that plumes were formed by sizable asteroids, which hit the Earth's surface in the recent geological past with force sufficient to penetrate the crust into the mantle. Since, the theory goes, the Earth's surface is largely covered by oceans and ocean crust is on the average one-third the thickness of continental crust, one would expect to find mantle plumes caused by such events more frequently below ocean crust than below continental crust.

Although there is no island chain in the region, Iceland is nevertheless believed to be located over a mantle plume and is — in this sense alone — like any other volcanic island. Iceland is just an accumulation of lavas on the ocean bed that has become sufficiently great for the upper surface to rise above the sea. But was the mantle plume beneath it the result of an asteroid collision? I asked Haukur Johannesson of the Museum of Natural History, Reykjavik, one of the country's leading geologists. He explained that, before Iceland existed, the southeastern coastal area of Greenland, the Faeroe Islands, and northern Scotland above the Great Glen were part of the same region of Pangea. When this area of the super-

THINGVELLIR, NEAR REYKJAVIK, SOUTHERN ICELAND

continent began to rift apart, the plume over which Iceland is now positioned was already in existence. If the plume had been caused by an asteroid, we could expect to find evidence of impact in the correlating regions of Greenland, the Faeroes, and northern Scotland, but none has been found. On the other hand there is the adequate evidence of the enormous lava fields that were formed when the rifting episode began. That Iceland was (and still is) at the center of a rifting process is demonstrated by the continuous series of ridges or swells on the seabed at right angles to the Mid-Atlantic Ridge, in a direct line between Greenland and

the Faeroes with Iceland equidistant between the two.

The general view of Icelandic geologists seems to be that the Earth's crust, normally rigid, has partially melted under Iceland and that above this exceptionally hot zone there are twenty-five or more volcanic magma chambers. The magma chambers are in line under the Reykjanes Ridge (the name given to the Icelandic section of the Mid-Atlantic Ridge) where it crosses the country from south to north. The actual process of rifting is not continuous but periodic, and the act of separation takes place in jerky movements as first one magma chamber inflates and erupts and then another. The *average* rate of divergence along the whole Mid-Atlantic Ridge

may be only a few centimeters a year, but individual magmatic events along the Reykjanes section of the ridge cause movement in the order of 1 to 4 meters in a period of a few days or weeks. A particular zone may erupt several times over a period of years and then stay dormant for several hundred years while another region of the ridge becomes active. It looks as if movement in one part of the ridge creates stress in dormant regions and as if accumulating stress is relieved only by corresponding movement in a previously dormant sector. Iceland's fissure zones, the magma chambers, and the Reykjanes Ridge in general are there-

Thingvellir is a long-dormant part of the Reykjanes Ridge. It is here that the Althing, the first Icelandic parliament and one of the earliest of all parliaments, met regularly in the midsummer of each year from about A.D. 930 until 1798.

But Thingvellir is also near part of a dormant fissure zone, 5 kilometers wide at this point. The fissure zone, or spreading ridge, is the flat area on the other side of the escarpment running diagonally from the center to the right of the panorama. The River Oxara is flowing through a minor fault caused by collapse of the fissure zone after activity had ceased and the basalt below the surface had solidified to a great depth. The collapsed surface was later covered by more lava, this time from a shield volcano, the low-profile peak in the center distance above the ridge. Most of the other volcanoes on the horizon on both sides of the ridge were formed under glaciers during the last Ice Age.

The well-drained and mineral-rich volcanic soils in Iceland, sometimes warmed by nearby hydrothermal activity, often support quite beautiful clusters and groupings of Alpine plants.

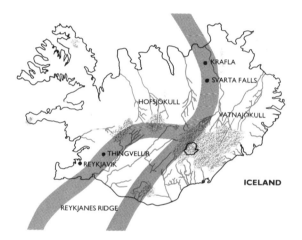

fore manifestations of the division of the ocean floor, not the cause of it. The most probable cause is the conveyor-belt movement induced by convection cells below the lithosphere (see page 2). Iceland and the Mid-Atlantic Ridge, of which it is a part, are simply above a region where convection cells interact.

Johannesson's explanations caused me to think again about the images of Earth's dynamics that I had been forming from general reading. I had got the idea that the dividing and spreading seabed pushes the continents along. Evidently this is not so. The two sections of the Atlantic seabed appear to be conveyor belts without their own power and therefore may not contribute to continental movement. As the lava from which the oceanic crust is formed is extruded from the central spreading ridge, the lava cools and solidifies to form a rigid floor of basalt. Everything on this floor is moved from east to west or from west to east, depending on which side of the ridge the object happens to be. Presumably the huge continents on either side of the Atlantic Ocean are moved by the same forces of convection under the Earth's lithosphere that drive the seabed conveyor belts. But as I was later to learn, other forces are also thought to play an important part in this process. From a human being's point of view the motion is incredibly slow, but each centimeter of movement per year translates into 10 kilometers in a million years; that's fast geologically.

The best known part of the Reykjanes Ridge is in the southwest of Iceland at Thingvellir, about an hour's drive northwest of Reykjavik. This sector of the ridge, about 5 kilometers from one side to the other, has been quiescent for centuries. As a consequence of long dormancy the last flow of basalt lava has had adequate time to cool and crystallize, and it has therefore done so to a great depth below the surface. As the enormous mass of basalt cooled, it contracted and sank lower in the Earth's crust, causing faults on both sides of the ridge valley. Rainwater and snowmelt streams run into the valley and collect at its lowest point to form a lake, Lake Thingvalla-vatn, at its southern end, where it is dammed by lava flows from another and more recently active fissure system. Although the surface of the lake, the largest in Iceland, is 114 meters above sea level, the bottom of it is 11 meters below sea level.

The geological term for a collapsed valley floor of the character of Thingvellir is *graben*, the German word for "ditch." The word *thingvellir* is Icelandic for "a place of assembly," and Thingvellir is in fact the original site of the Althing. The Althing is generally assumed to have been founded around A.D. 930 and to have continued with annual midsummer meetings at Thingvellir until 1798; half a century later it was reestablished in Reykjavik. Like all the early parliamentary systems in Europe the original Althing was more an aristocratic than a democratic assembly, where the principal landowners made laws and punished those who broke them, as the drowning pool near the site of the Althing testifies.

Photographs of Thingvellir appear in many geological books, and I had gained the impression from them that the fault through which the River Oxara flows was indeed the Thingvellir fissure zone, but this was not the case. The fissure zone is on the plain beyond the edge of the fault escarpment. The fact that the zone is kilometers wide enhanced my general sense of awe as I saw the site for the first time. On this scale the eruptions of fissures must be spectacular. Fortunately I had an introduction to the film producer Vilhjalmur Knudson and his American wife, Lynn, who live just across the street from the Holt Hotel in Reykjavik, where my wife, Joy, and I were staying. In a studio connected to their house they run a small cinema which is always crammed to the doors for every performance of their films. For me the most interesting scenes in Villi's films were those of eruptions of the Krafla fissures at the northern end of the Reykjanes Ridge. Seeing these events recorded on film immediately following a visit to dormant Thingvellir brought that scene to vivid life. The magnitude, quite apart from the magnifi-

cence, of fissures in eruption seemed overwhelming.

The Krafla eruptions Villi Knudson had filmed were part of a series that began in 1975 and have continued intermittently ever since. The previous eruptions at Krafla were in the period 1724 to 1729. Geophysicists cannot predict when and where such a series is likely to commence in a dormant zone like Thingvellir. But once a magma chamber in a particular region begins to stir and to inflate with magma from the mantle, the physicists can predict a likely date. But to forecast the day or the week or even the month with any certainty is not feasible. For this reason filming such eruptions demands constant readiness once a forecast has been made. Patience is sometimes stretched to months of tense waiting. Krafla, which is very remote from Reykjavik, had been forecast to erupt but was long overdue. Villi had been poised expectantly. It might just be my good luck to be there at the right time, he suggested with a grin.

Our objective was simple enough: to drive up the Reykjanes Ridge from south to north across the uninhabited interior of Iceland and to visit the Krafla fissure zone at the far end. Joy and I equipped a Range Rover and left Reykjavik early one morning in pouring rain.

Part of the tremendous charm of Iceland is the fact that, as the limited paved highways near Reykjavik give way to lava-gravel surfaces, there is an immediate transformation from being more or less anywhere to being somewhere extraordinary. Every now and again packs of Icelandic ponies, led by the owner sitting astride one of his animals, appear unexpectedly around bends in the road. A friendly wave from the rider and then those

wonderful animals trot gaily by. One simply has to stop and admire them: I'm sure that Icelandic ponies come straight out of a book of Norman Thelwell drawings. Perhaps Thelwell invented them?

The gravel road deteriorates into a 250-kilometer-long track across the country as one progresses toward the wet and forbidding interior. Perhaps we were just unlucky, but it seemed to rain perpetually. The first difficulties proved to be only decisions about which of two — and sometimes three — diverging forks to follow. Fortunately we guessed right and finished that day's run off the track on the brow of a smoothly undulating hill overlooking Hofsjökull, the second largest ice cap in Iceland. It was nearly midnight; and because the ice cap was due west of us and it was midsummer near the Arctic Circle, a prolonged sunset occasionally penetrated the cloud cover that hung over the cap. From our elevated position Hofsjökull looked huge. It is 1,765 meters in height and stretches as far as one can see to left and right. Toward its near edge it is studded with volcanoes, a few of which are conical in shape; most, those nearest the edge of the ice cap, are table mountains, flat-topped volcanoes formed under the ice-cap and now being revealed as the ice recedes. Recession is a common fate of glaciers all over the Northern Hemisphere.

The landscape now became undulating, with low-profile hills, barren of trees and vegetation, and with table mountains to the east and to the west on either side of the ridge. Ridge is perhaps a misnomer in the usual sense

The panorama is of a beach composed of hyaloclastites formed from exploding lava under the sea. The basalt formation in the left foreground is an eroded remnant of lava that so-

lidified in a volcanic vent without reaching the sea. The smaller picture is of pillow lavas formed just below the critical depth and therefore of open texture.

The black hyaloclastite beach at Dyrholaey, on the southeast coast of

Iceland, attracts thousands of seabirds to rear their young on the surrounding basalt cliffs. One of the attractions is that a black lava beach more readily absorbs heat from the sun than a

light-colored sandy beach does. Flocks of birds can be seen sunning themselves in the center foreground.

of the word; like Thingvellir it took the form of a wide, flat valley along its length. I found it difficult to resist associating the scene with my idea of the seabed's appearance. After all Reykjanes Ridge should be at the bottom of the North Atlantic. Table-topped volcanoes in some ways resemble the oceanic crust because of the extraordinary way in which they were formed under the ice cap. Their summit surfaces are often studded with smaller cones. I wondered what processes went on under the ice during volcanic eruption to cause these peculiarities. Later I got an explanation from a volcanologist.

It appears that about 10,000 years ago, at the end of the last glacial advance, two-thirds of Iceland was covered by thick ice, perhaps as much as 2,000 meters deep in parts. Ice covered the Reykjanes Ridge for much of its length. The same processes of seafloor spreading operated then as now with the difference that fissure eruptions took place under the ice cap. Volcanic eruption under ice can melt 1 cubic kilometer of ice within a few days, a process that causes a deep lake to form. Continued eruption into the subglacial lake that forms under the cap causes the erupting lava

to form pillows beneath its surface, the same type of basaltic lava formation associated with the Mid-Atlantic Ridge and other spreading ridges on the seabed. As eruptions continue, more ice melts around and above the growing pile of pillows until the pile begins to collapse under its own weight, spreading out to form the base of a flat-topped volcano. The top of the pile will eventually accumulate to reach a level below the lake surface at which water pressure is insufficient to prevent the lava from exploding (instead of forming pillows). At this critical point the explosions form vast quantities of glassy granules, hyaloclastites, which accumulate to envelop the underlying pillows. The hyaloclastites then consolidate beneath the level of the subglacial lake to complete the formation of a flat-topped table mountain. When eruptions continue beyond this point, as they often do when shield volcanoes erupt beneath the ice, a flat cap of lava forms on the surface of the accumulated lava sheets above the water level with a volcanic vent somewhere on its top surface — which explains the peculiar-looking cones on some of the table tops.

It was pouring rain again the following morning, after our overnight stay, and when we left, none of Hofjökull's table mountains

DYRHOLAEY, SOUTHEAST COAST OF ICELAND

SVARTA FALLS, ISHOLVATN, NORTHERN ICELAND

This panorama was taken in the northern section of the Reykjanes Ridge in Iceland. The ridge is part of the mid-Atlantic ocean-floor spreading zone that rises above the surface of the sea, crosses Iceland from south to north, and then descends into the sea again at the latitude of the Arctic Circle.

The picture illustrates an old lava flow in the rift valley that has been cut by the River Svarta. During eruption the top and bottom sections of this lava stream cooled and solidified while the central section remained semiliquid and cooled slowly — slowly enough to allow polygonal columns of basalt to form while the top and the bottom sections remained contorted.

was visible. The flat undulating hills over which we were traveling looked even more like a seabed. We came to the midpoint in the crossing and found there an overnight hut open during the summer months. This is used mainly by passengers of the extraordinary bus that daily traverses the island from Reykjavik to Akureyri, a distance of about 450 kilometers. The bus looks rather like a water beetle: small elongate body set high off the ground above very large, heavily tired wheels. It is designed not only for the rough terrain but also to negotiate rivers in flood. Beyond the hut we came to our first river in flood. It looked most uninviting, very wide and fast flowing. The river was one of a number of such rivers, which flow from a remnant ice cap of the Ice Age called the Vatnajökull, the largest in Iceland, which was out of sight in the rain clouds to the east. Having first used binoculars to get a line on the tread marks on the opposite side, I selected four-wheel drive and the lowest gear available and took the plunge. Slowly, carefully ignoring the torrent and keeping my foot well away from the clutch, I gingerly persuaded the Range Rover across. (Perhaps it would be more truthful to say that the Range Rover gingerly persuaded me across.) Later, having

navigated numerous other rivers in far greater turmoil, I wondered why I had been so timid.

The bare submarine landscape was suddenly relieved by a green oasis that loomed out of the pouring rain in ghostly fashion. Unexpected low-lying succulent plants shared ground with patches of pink flowers. Bright green-yellow moss mingled with the snow patches that hung from the low eastern slopes of a neighboring mountain shrouded in clouds. The whole eerie scene was interspersed with mounds of black volcanic ash; obviously the greenery resulted from nearby hydrothermal activity. Out of the mist a lonely walker appeared ahead, hunched beneath an enormous backpack. I stopped, but he passed without acknowledgment. So wet was he, I imagined, that he must be cursing the day he set out on his marathon and seething with resentment at the thought of anyone warm and dry in a motor vehicle. The oasis passed, desolation intensified, and swollen-river crossings multiplied. The prospect looked grim as there were many more rivers to cross. After consultation with my codriver I decided to take a short cut to our overnight destination, the Svarta Falls. We began to climb steeply as the secondary track crossed challengingly over a mountain range. The mist deteriorated into dense fog.

The Svarta Falls in the rain were impressive but somber with a funereal black surround of wet basalt rock. I wanted particularly to photograph a good cross section of crystallized basalt because I knew this rock to be the characteristic material of the ocean crust and therefore representative of 70 percent of the Earth's surface. Most of the lava cut by the river in this region is the result of one eruptive event. The lava on the top of the flow and on the bottom of the flow had congealed rapidly because of contact with cooler environments, the top surface with the air above it and the bottom surface with the comparatively cool, old rock surface on which it was flowing. The center of the lava sheet had remained relatively fluid for a long time after the general flow had ceased. It cooled slowly enough to crystallize into the hexagonal columns characteristic of columnar basalt.

This basalt section cut by the River Svarta is analogous to the structure of the ocean crust without its cover of sediments. Geophysicists believe that the upper 3 kilometers of the ocean crust, roughly level with the magma chambers of the Mid-Atlantic Ridge, consist of a 2-kilometer thickness of vertically sheeted dikes covered by about a 1-kilometer accumulation of pillow lava. Below the sheeted dikes the 2.5-kilometer lower stratum makes contact with the very hot upper part of the mantle. The total thickness of the ocean crust is therefore believed to be about 5.5 kilometers. Imagine this section of the Svarta Falls lava flow to be a rough model of such a structure. The upper layer of congealed lava could represent the pillow lavas; the columnar basalt could represent the sheeted dike; and the bottom layer of congealed lava could represent the basic material in contact with the upper mantle.

This model is of course composed of cold static material. The real ocean crust is very hot toward the Mid-Atlantic Ridge and becomes cooler as it progresses outward from the ridge. The expanded hot basalt is lighter than

This is an imaginary section through the Mid-Atlantic Ridge but the scale is deceptive. The ocean crust averages 5.5 kilometers in thickness. At the spreading ridge from which the continents diverge the thickness is minimal because it is at this point that ocean crust is actually being created.

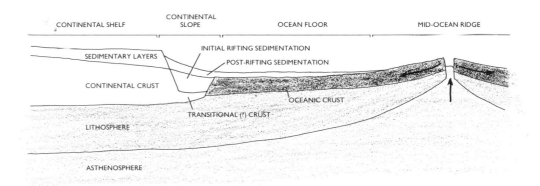

This sketch illustrates the assumed relationship of the passive margin between the North Atlantic ocean crust and the North American continental crust (together called the lithosphere) and between the semiplastic portion of the upper mantle (the asthenosphere) and the Mid-Atlantic Ridge. The ocean bed is raised toward the rifting ridge because the heat of magma near the surface at this point causes expansion. Sediments on the ocean bed increase in depth as they near the edge of the continent. Sediments from the continent form basins, which depress the lithosphere.

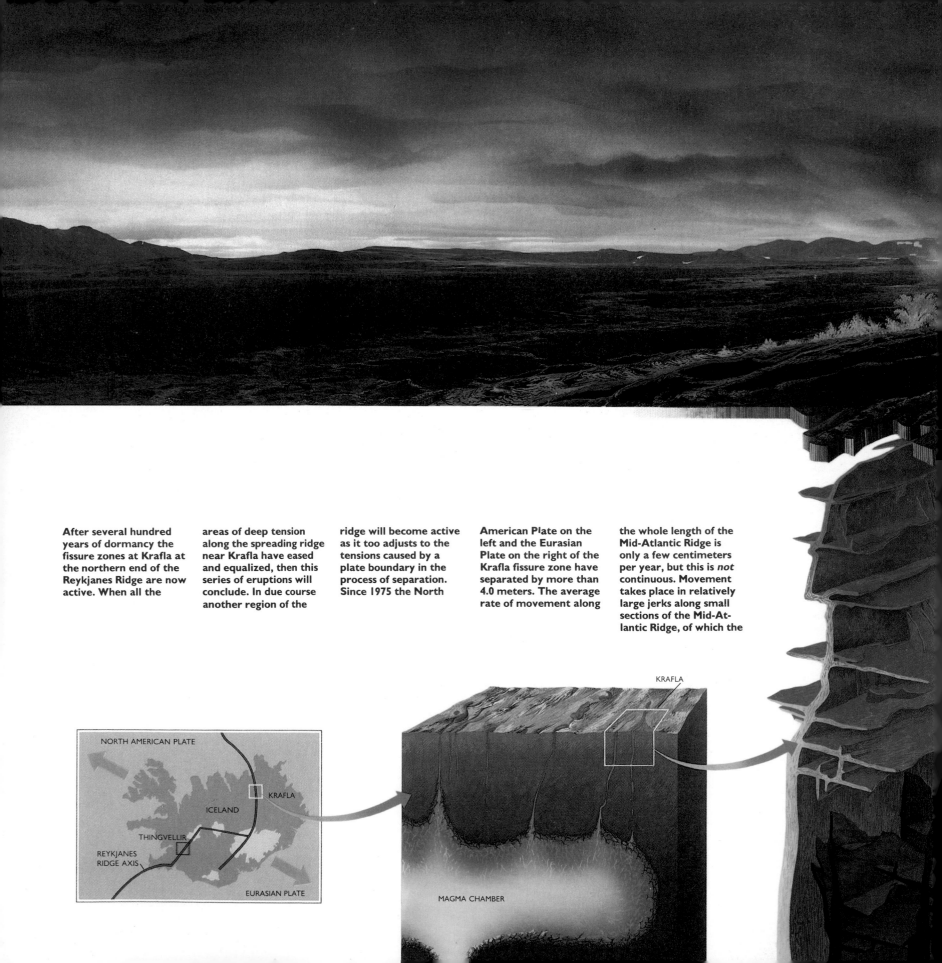

After several hundred years of dormancy the fissure zones at Krafla at the northern end of the Reykjanes Ridge are now active. When all the areas of deep tension along the spreading ridge near Krafla have eased and equalized, then this series of eruptions will conclude. In due course another region of the ridge will become active as it too adjusts to the tensions caused by a plate boundary in the process of separation. Since 1975 the North American Plate on the left and the Eurasian Plate on the right of the Krafla fissure zone have separated by more than 4.0 meters. The average rate of movement along the whole length of the Mid-Atlantic Ridge is only a few centimeters per year, but this is *not* continuous. Movement takes place in relatively large jerks along small sections of the Mid-Atlantic Ridge, of which the

NORTH AMERICAN PLATE

KRAFLA

ICELAND

THINGVELLIR

REYKJANES RIDGE AXIS

EURASIAN PLATE

KRAFLA

MAGMA CHAMBER

Krafla area is thought to be typical.

One of these events, based on the present series at Krafla, has been reconstructed here over the author's panorama taken from the center of the fissure zone. The overall scene is set in a volcanic caldera, with a magma chamber beneath that inflates and deflates according to the pressures of the magma within it.

the consolidated cold basalt, and this is the reason why the regions on either side of the ocean ridge are higher in elevation than are the neighboring ocean beds. The old basalt lava flow at Svarta gives a sense of the structure of the ocean crust but no idea of its scale. The thickness of the ocean crust is equivalent to 3.5 times the depth of Grand Canyon at its deepest point. Some conveyor belt!

In the far distance up the rift valley beyond the Svarta Falls I had noticed a small white building with a red roof. It proved to be a farmhouse, the first habitation after our crossing of the interior. The Krafla caldera and the fissures that Villi Knudson was poised to film were only a few hours away. When we reached the region of Krafla, I found that the route that Villi had marked on my map led into a maze of steep, elongate hills. I suddenly realized that these were part of a series of volcanic vents, a fissure swarm, just south of the central caldera area. Each hill was made up of sandy material ejected from fissures along the crest. After one or two false starts we eventually found our way through the fissures onto a flat grassy surface running beside a lava field. Large chunks of lava towered above the vehicle, and we had to stop occasionally to climb them for a survey of the route. I was making for a small volcano surrounded by the lava flows of the last eruption. It seemed to command the kind of view I was looking for so that I could take a panorama of the main fissure zone for a reconstructed cross section of the caldera.

The grass ran out, and we found ourselves facing a wall of lava with an opening presumably cut to provide access for the service vehicles used during the construction of a power line. What a power line was doing running straight through an active volcanic field I could not guess. There followed perhaps one of the most difficult jeep trails that I had ever negotiated: short, steep turns at sometimes alarming angles, over very sharp chunks of black lava that threatened to cut the tires to shreds. After this effort the small volcanic vent fortunately proved to be a good choice.

We carried equipment up the steep, unconsolidated, and sometimes precarious slope to the summit, where the surface was distinctly warm to the touch. Photography finished, we were able to sit down and look at that quite remarkable scene at leisure.

The Krafla volcanic basin is an active rifting zone on the axis of the Mid-Atlantic Ridge. Nowhere else on Earth is there such a considerable length of active spreading above the level of the sea. The top of that small volcano seemed a good place to think through what I had been told about volcanoes. Volcanic systems in the upper crust, so I understood, are fed by reservoirs of melted material from sources deep below the lithosphere; these reservoirs are called "magma chambers." In the case of the Krafla caldera the associated magma chamber is situated about 3 kilometers below the surface and was dormant for about 200 years before the current series of volcanic events began in 1975. During the period of dormancy sections of the Mid-Atlantic Ridge to the north and south of Krafla had moved apart, thus creating tension in the Krafla region along a zone several hundred kilometers in length. The relief of this tension has resulted in a 3- to 4-meter separation of east from west in the Krafla area since 1975.

During each of the current series of volcanic pulses that have resulted in this considerable movement the now-reactivated caldera inflates like a balloon, up to 10 millimeters a day as magma moves into the chamber, and it was inflating as I was sitting there thinking about it. When pressure within the chamber reaches a critical point, fissures below ground are forced open by intruding magma. The fissures link together to form continuous dikes to the north and south along the ridge axis. If dike formation is sufficient to relieve pressure in the magma chamber, the caldera as a whole deflates and the volcanic pulse comes to an end; no surface eruptions occur. But if the dike swarms themselves are not sufficient to relieve the pressure in the chamber, magma escapes in spectacular fountains on the surface of the caldera. Such surface eruptions then continue until pressure is relieved in the magma chamber; deflation follows. After deflation the fissures beneath the caldera's surface close, and for reasons not yet explained other fissures begin to open at distances up to 80 kilometers away to form new volcanic

dikes. These episodes are accompanied by earthquake tremors, which cause vertical ground movement of up to 2 meters as far north as Axarfjördur, near the Arctic Circle, 90 kilometers away.

As Hauker Johannesson had explained to me, the activity of magma chambers beneath the Reykjanes and other sections of the Mid-Atlantic Ridge certainly makes manifest the movement of North America away from Europe; but the activity is the result of such movement, not the cause of it. With that Svarta Falls model of the ocean crust in mind and the drama of Krafla spread out below, I just sat and stared for a while at the plate boundary between North America and Europe.

We had become very cold while we were perched on the top of that small volcanic cone even though clad in down clothing in midsummer. We made our way back through the lava field to find a sheltered spot for the night among the fissure-swarm hills. Halfway through an evening meal there was a quite strong earth tremor. Not a word was said. We quietly packed up and left.

My next objective was to visit continental regions on opposite sides of the plate boundary, regions which had once been part of the same supercontinent, Pangea, but which were now widely separated. F. W. Dunning, Curator of the Geological Museum in London, had suggested specific places in North Wales and Newfoundland that were believed to have such a correlation.

North Wales is an evocative place for me — a place of childhood memories and holidays with my Welsh-speaking parents, of calling to sheep dogs with the few Welsh words I knew and getting a response, of calling again in English with no response at all. It tickled me to think that the dogs couldn't speak English. Much later, memories of rock-climbing ventures with Joy on Milestone Buttress above Telford's historic Holyhead road before tackling the more serious climbing routes on Hollytree Wall and in the Devil's Kitchen above Llyn Idwal. Then the real test on the cliffs of Lliwedd on Snowdon's flanks. Later again,

CWM NANT COL, NORTH WALES

Geologists believe that the Harlech Dome in North Wales and part of the Avalon Peninsula in Newfoundland were formed during the same period of time when they were both part of the same continent. Today North Wales and southeastern Newfoundland are 2,200 kilometers apart but equidistant from the Mid-Atlantic Ridge that separates them.

The rocks of Cwm Nant Col pictured here are part of the Harlech Dome. They date from 570 million years to 500 million years ago, and most of them were formed as sediments in shallow seas. They are different from neighboring Snowdon and Cader Idris rocks, which are much younger formations.

There are certain features in the Harlech region that correspond exactly to those found on the Avalon Peninsula of Newfoundland. One matching formation in both countries contains cube-like crystals of almost pure manganese, which were precipitated on the bottom of a quiet freshwater lagoon. Another rock formation in this picture contains fossil trilobites that are the same as those found in similar rocks in Newfoundland.

PARADOXIDES

This fossil trilobite, *Paradoxides,* is an important key to the identification of rocks that were formed at the same time on the same landmass, a landmass that is now separated by the Atlantic Ocean.

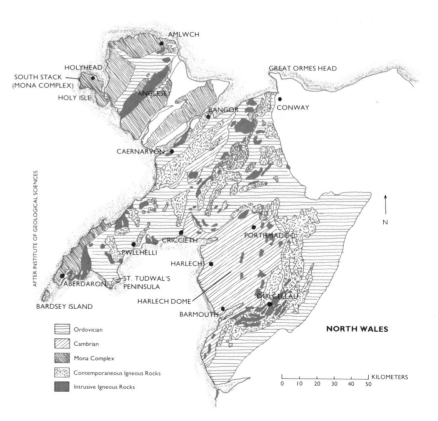

AFTER INSTITUTE OF GEOLOGICAL SCIENCES

AMLWCH
HOLYHEAD
SOUTH STACK
(MONA COMPLEX)
HOLY ISLE
ANGLESEY
GREAT ORMES HEAD
BANGOR
CONWAY
CAERNARVON
CRICCIETH
PORTHMADOG
PWLLHELLI
HARLECH
ABERDARON
ST. TUDWAL'S
PENINSULA
BARDSEY ISLAND
HARLECH DOME
DOLGELLAU
BARMOUTH

NORTH WALES

N

Ordovician
Cambrian
Mona Complex
Contemporaneous Igneous Rocks
Intrusive Igneous Rocks

KILOMETERS
0 10 20 30 40 50

with a young family, across the Menai Straits to the marvelous sandy beaches of Anglesey. All those years ago these places were attractive rocks, mountains, and beaches. Now, no stranger, I was curious to see the Harlech Dome region with a very different eye.

The Harlech Dome is an oval-shaped mountainous area in the neighborhood of Harlech Castle, on Tremadog Bay and the Barmouth Estuary, on the west coast of North Wales. On its landward perimeter it is bounded by the Snowdon and Cader Idris mountain ranges, which are mainly volcanic in origin, while the Harlech Dome itself consists of mountains formed mainly from far older sedimentary rocks. The sediments that formed the dome began to accumulate about 570 million years ago. Each stage of accumulation reflected the changing environments from which the sediments originated: shallow seas, brackish freshwater lagoons, and estuaries. Later these rock formations were disrupted by continental movements, which caused them to be uplifted, folded, compressed, and intruded with magmatic rock. About 60 million years ago, after a long, scientifically speculative, history, the region was broken up — rifted apart by what is now the Mid-Atlantic Ridge. The Harlech Dome was part of the eastward-moving block and the Avalon Peninsula of Newfoundland was part of the westward-moving block.

I had read about these dramatic events, and I had been told about them; but rather than just accepting these statements without question, I felt that it would be an interesting experience to follow a few of the clues that had led geologists to such mind-boggling conclusions. To me it seemed quite astonishing that the Harlech Dome and Harlech Castle and Rhinog Fach of my unsuspecting boyhood should have shared a continent with a large chunk of Newfoundland now 2,200 kilometers away.

One of the oldest sedimentary layers in the Harlech Dome was laid down in an extensive, shallow, brackish lagoon fed by streams of water rich in manganese. Evaporation in the lagoon caused the precipitation of manganese crystals, and these settled into the muddy bottom, which eventually turned into shale, manganiferous shale. The best exposure of this formation is in Cwm Nant Col, in the very heart of Harlech Dome, where it was mined until recent times. Joy and I clambered up the tailings of the old mine shafts on the flanks of Rhinog Fach, looking for chunks of discarded rock that might contain the cubical crystals of manganese. We broke open rock after rock until Joy eventually found a matrix glistening with crystals — the first time they had been exposed to the light of day since they were formed about 500 million years ago. We collected samples of the crystals and the shale to compare with those which, so I had been told, were to be found in Newfoundland.

Manganiferous shales would present one sort of evidence for the relationship of the Harlech Dome to the Avalon Peninsula. Was there another? Fred Dunning had suggested trilobites of the genus *Paradoxides*, which appear in a Harlech Dome stratum perhaps 30 million years after the manganiferous shales. Arthropods, the general name for trilobites and their descendants, were the earliest form of hard-shelled sea creature on Earth, and trilobites were one of the most successful biological groups ever to evolve. During the 300 million years that trilobites appear in the fossil record they developed into hundreds of varieties, each with different characteristics of size, shape, and habitat. When trilobites died, their hard parts fossilized in the sediments at the bottom of shallow seas. A gradual peppering of these exoskeletons on the seabed created horizons that marked the first and the last appearance of some species. The more specialized the trilobite, the greater its physical modification to suit a particular environment. The shorter the distance in time between its horizons, the more useful the fossil becomes for indexing a specific period in time. *Paradoxides* is one such key fossil; for it lived only in one quite extensive area for a short period of time. Any rocks containing *Paradoxides* are therefore related in time and place.

It was still raining when a few weeks later the hunt for *Paradoxides* was resumed in Newfoundland, 2,200 kilometers west of the Harlech Dome. The translation from North Wales to Newfoundland was remarkable because of the similarities rather than the differences: the people, the houses, the character of roadside villages, the topography, the large number of quarries, and the wet summer weather; it could have been North Wales. Geologists of the Newfoundland Survey in St. John's had directed me to a place near Topsail, 25 kilometers away on the other side of the Avalon Peninsula, and there I grubbed about in muddy loose shale looking for trilobites. Suddenly I found a pocket of fossils in the shale, then another, and another — a veritable graveyard. They were, so it was later confirmed, the same species of *Paradoxides* found in the Harlech Dome. The Newfoundland geologists also confirmed that the manganiferous shale samples I had brought matched those of the Avalon Peninsula. But there were two more facts to add to my modest checkup on geological science. The first was the Mona Complex. This very ancient basement rock, which underlies parts of North Wales (and certain other regions of the British Isles) and which outcrops in spectacular fashion near Holyhead, Anglesey, also underlies the Avalon Peninsula with nothing but the broad reaches of the North Atlantic between. And the second fact: that fracture zone astride the displaced Mid-Atlantic Ridge, which I mentioned at the beginning of this chapter. Its center lies exactly halfway between the two countries in the middle of the North Atlantic. All that was needed to complete my modest investigation was to add QED.

Paradoxides had been extinct for over 400 million years before the Harlech-Avalon rock formations began to drift apart. During that immense period of time hundreds of thousands of species of land-based flora and fauna, including the early primates, had evolved. Very much later they had become separated from each other as the supercontinent of Pangea began to break up. The separated primates became the New World Monkeys and the Old World Monkeys, new species from common stock. But the hominids had not evolved at the time of the separation of Africa from the Americas, and they later developed into *Homo habilis*. There had been no radical speciation of the kind forced upon other animals by continental division. Continental distribution of mankind was achieved by migration. I wondered when man had reached Newfoundland, the most southeasterly point of the North American continent. (The most northeasterly point is Crown Prince Christianland in Greenland.)

I had heard of a quarry with extraordinary carvings on its walls somewhere in northeastern Newfoundland. "Yes," a geologist told me, "there is a strange place near the Baie Verte Fault, the old Laurentian continental margin, a soapstone quarry, I think, used by the Vikings. Place called Fleur de Lys. Go as far the road will take you in the village and ask for the carvings." It was a long haul from St. John's.

On our arrival the first discovery was that I couldn't understand a word of the English spoken at Fleur de Lys, which proved to be a small fishing village at the end of a remote peninsula. The language was a strange mix of British dialects, with perhaps overtones of local modification over the centuries; these people must be directly descended from some of the earliest permanent settlers in North America. Later I learned that the mixed dialect of Fleur de Lys was derived from the speech of the west of England three centuries ago and made use of idiom and pronunciation now forgotten in Dorset, Devon, and Somerset. Joy came to my rescue; she seemed to understand the dialect better than I did.

We were directed away from the sea cove at the end of the road toward a small house sheltered by a cliff. We followed a path between the house and a vegetable garden and came upon the most astonishing sight. The lower part of a stretch of cliff perhaps 15 meters high, at the rear of the dwelling, was marked with excavated hollows in orderly lines. Some appeared as if cut by an oversized, round, ice-cream scoop; others had a raised center. They were approximately the same overall size, about 35 centimeters in diameter. One section of cliff had been cut away in an overhang so that the carvers must have worked from some kind of platform. Other carvings on other sections of cliff were in even more inaccessible places. The working faces covered the cliff for 50 meters or more. I wondered whether the Vikings had in fact been responsible for this unique quarry.

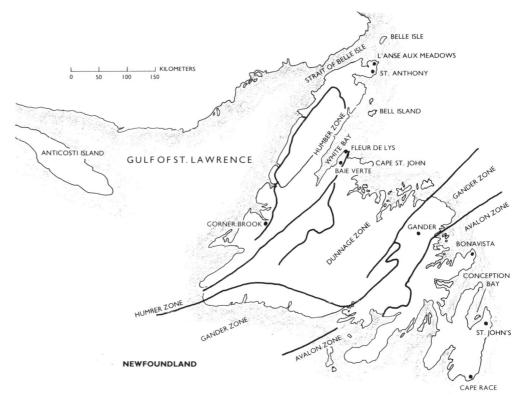

NEWFOUNDLAND

That night I called James A. Tuck, an archeologist at Memorial University in St John's, and an authority on aboriginal history in Newfoundland and Labrador. "What on earth are these carvings at Fleur de Lys? Are they Norse?" Tuck chuckled. They were cut, he told me, by Paleo-Eskimos sometime between 1,500 and 2,000 years ago. These aboriginal people probably believed that to carve their artifacts from "living rock" gave the vessels that they shaped from the lumps qualities superior to those shaped from "dead rock" picked up from the base of the cliff. The cliff is of soapstone, a sedimentary rock that has been cooked and therefore metamorphosed by deep burial and pressure. Its main constituents are talc and mica, which impart a soft silky feel to its polished surface. Soapstone is relatively easy to cut with a harder rock, such as a flint tool napped from chert. There are modern talc (and asbestos) quarries between Fleur de Lys and Baie Verte.

It was surprising to discover from Jim Tuck that the Paleo-Eskimos had been preceded in Newfoundland by other early Inuit, people of the Dorset Culture, back to 3,000 years before the present. (The term Inuit is preferred by the Eskimo people: Both descriptions will be used in this book as context dictates.) The Dorset Eskimo were preceded by the Maritime Archaic Indians, who date back to 5,000 years before the present. In fact Archaic Indian sites dated to 9,000 years before the present have been discovered on the Labrador coast near the Strait of Belle Isle, which separates the mainland from Newfoundland. I wondered whether Newfoundland had been the ultimate meeting place of the people who had migrated from Asia to North America, some via the Arctic (the Eskimo) and some via continental North America (the Indians)? Whatever the answer, finds of notched double-sided projectile tips, probably arrowheads, of a type originating with the Maritime Archaic Indians, together with other artifacts resembling Dorset Culture prototypes link these very early

SOAPSTONE QUARRY, FLEUR DE LYS, NORTHEASTERN NEWFOUNDLAND

The extraordinary scoops in the soapstone rock face at Fleur de Lys, Newfoundland, were carved by Paleo-Eskimos between 1,500 and 2,000 years ago. These ancient people believed that bowls or drinking vessels cut from the living rock had special properties, which dead rock picked up at the base of the cliff did not possess. An occasional scoop has a raised flat surface in the center of the depression from which the carving was cut — no doubt to form a flat base.

Harebells and fireweed grow in profusion in the vicinity of the quarry. But there was no sign of the heraldic lily from which the locality had got its name. The villagers speak a dialect derived from 16th-century English, and many of them are direct descendants of the original immigrants.

people with the comparatively modern Beothucks, the Indians who gave the expression Red Indians to the English language. The Beothucks, now extinct, were named Red Indians by 16th-century explorers and settlers of Newfoundland because of the liberal use of red ocher on their bodies. It is also probable that it was the Beothucks who were referred to as *skraelings* ("natives") by the first Europeans known to have settled in Newfoundland, the Norse at L'Anse aux Meadows circa A.D. 1000.

L'Anse aux Meadows is on shallow, sheltered, Epaves Bay overlooking the Strait of Belle Isle on the far northern tip of Newfoundland. It is just south of the mainstream of pack ice that drifts down the east coast of Greenland into the Atlantic during the summer months. Icebergs from the glaciers on the west coast of Greenland regularly sail through the strait in majestic fashion, and some become stranded in shallower water if wind drives them too close to shore. The Norse site at L'Anse aux Meadows was discovered in 1960 by the Norwegian explorer and writer Helge Ingstad and was later excavated by his wife, Ann Stine Ingstad, in company with other distinguished archeologists from Canada, the U.S., Scandinavia, and Iceland. Helge Ingstad had reviewed the Icelandic sagas and had concluded, after searching the coastlines of New England, Nova Scotia, and Labrador, that this area of Newfoundland best fitted the old Norse descriptions of the mysterious and still highly controversial Vinland.

Before leaving Iceland, I had visited the Arnamagneum Institute in Reykjavik to see on display the wonderfully embellished vellum pages of some of the sagas, which were produced from the 13th to the 18th centuries, and then to discuss their significance with Dr. Halldorsson, the scholar responsible for the interpretation of the *Greenlanders' Saga* and the *Saga of Eric the Red*, perhaps two of the most significant documents in the early European history of the North American continent.

L'ANSE AUX MEADOWS, EPAVES BAY, NEWFOUNDLAND

By the year A.D. 1000 a Norse settlement had been established on the shore of a shallow sheltered bay in the most northerly part of Newfoundland. The site, which was abandoned after a few years' use, was rediscovered in 1960 by Helge Ingstad, a Norwegian explorer-writer. Ingstad had been searching for "Vinland," one of the several Norse discoveries described in two ancient Icelandic sagas.

The excavations of the eight structures built on the site and subsequently re-covered with turf can be seen in the panorama. They encircle an embayment that is now raised several meters above present sea level because of the continued rebound of the Earth's surface from the weight of ice that depressed it during the last Ice Age. The excavations brought to light adequate proof of Norse occupation, including an Icelandic lamp and the fire-cracked rocks of a sauna hut. Carbon 14 dating of the charcoal used for smelting iron and for domestic purposes established the probable dates of occupation.

The interior is of a reconstructed Norse longhouse: The reconstruction conforms to the plan of a structure on one site that was excavated and is now re-covered with turf.

The white bridge just visible in the middle of the panorama crosses a stream near the site of the smithy.

The remnants of four cairns, two of which were reconstructed by Ingstad, were found on the top of the raised area that forms the horizon of the site panorama. The purpose of these cairns is unknown. Perhaps they were sundials or navigational aids; but whatever their purpose, they closely resemble similar Norse constructions in Greenland and elsewhere.

Cloudberries, whinberries, and other edible berries grow in profusion in the bog near the cairns over the brow of the hill, along with cotton grass, which was used for lamp wicks.

Saga is an old Norse word from the root "to say" and is used in the sense "a spoken story." There are many sagas: about the kings and bishops of the day in Norway, Iceland, and Greenland, about the Norse people, and about events. The stories were passed down partly by word of mouth, partly by parchment documents, and partly on stylus-engraved metal or stone. No doubt truth became distorted and embellished during this process because the custom was to use fable and allegory. But no judgment should be made about the relevance of the *Greenlanders'* and *Eric the Red* sagas without first taking into account that at the time of the discovery of Vinland, Markland, and Hellalund — now broadly presumed to have been Newfoundland, Labrador, and Baffinland — a serious attempt to colonize Greenland was in progress and that it continued successfully for 500 hundred years before it ended in mysterious circumstances around A.D. 1500: The colony just ceased to exist without record or certain explanation. There were as many as five or six thousand Norse settlers in Greenland at different times during this period; so this was not a casual excursion.

The two sagas that relate the story of the Vinland voyages were inscribed on parchment in Iceland about 200 years after the event and are believed to have been produced quite independently. They can best be termed folklore, and how accurate they are is indeed problematic. But the accounts of aboriginals in Vinland cannot have been invented, and the descriptions of the landscape in both sagas are too similar and detailed to have been coincidental or fictitious. Halldorsson provided me

with a summary of descriptions of the Vinland landscape that he had made from both sagas. A selection of these, translated from Icelandic, accurately describes the country in the vicinity of L'Anse aux Meadows: "The land is not mountainous, but wooded with low hills"; "An island lay north of the land, with a sound between it and a headland jutting out to the north"; "West of the headland jutting north out of Vinland the sea was very shallow at ebb tide"; and so on.

Norse longhouses and other structures, eight in all, have been excavated on the L'Anse aux Meadows site discovered by Helge Ingstad. The houses had been built around an old embayment at a level just above what would have been the high-tide mark at that time. The surrounding land has rebounded from the weight of ice once upon it and is now raised well above the old embayment: Rebound continues for thousands of years after the ice sheet that depressed the terrain has melted. The now-raised beach level is one of a number of factors vouching for the antiquity of the site. But most important are the carbon 14 datings of charcoal recovered from fireplaces and the smithy in which bog iron collected from the site was smelted. All provide dates in the vicinity of A.D. 1000. Bone needles, iron rivets, nails, iron slag, smelted copper, perforated glass beads, a soapstone spindle, shaped pieces of wood, and most significantly an Icelandic lamp — all lend testimony to a Norse presence, but none so poignantly as the fire-cracked rocks of a sauna hut.

Beyond the fact that some buildings were destroyed by fire and that it is unlikely that the site was occupied by the Norse for more than a few years, the fate of this first European community in Newfoundland is not known. Similarly the fate of the formidable Norse settlements in Greenland, very much part of the North American continent, is not known.

Today's continents were within comfortable walking distance of their present geographic positions by the time the first tool-making hominids had evolved on the African continent more than 2 million years ago. The Atlantic Ocean was almost as broad then as it is now, and Newfoundland had already become the farthest southeasterly point on the North American continent. It was therefore a most significant moment for mankind when the Beothucks and the Inuit met the Norse in the region of L'Anse aux Meadows around A.D. 1000.

The Norse had broken the unwritten rule of evolution. They had cut across the protective barrier of an ocean and so had avoided the process of adaptation, the process that thus far had ensured the survival of aboriginal man. They had to adapt promptly or die or leave. Centuries later other Europeans began to arrive in North America. They too very nearly failed in their attempts to colonize. They owed the slender means of their survival to aboriginal technology, to the very people they inadvertently decimated with European diseases. The newcomers later realized that the North American environment was fundamentally different from that of Europe. They discovered differences in scale and character of the landscape, in scale of natural disasters, in natural resources, in patterns of weather, and in climate — differences fundamental to the anatomy of the continent. Only when they made these discoveries and adapted to them did the newcomers become Americans. Newfoundland was the starting place of my venture to discover the nature of that anatomy — the making of a continent.

THE NUCLEUS

THE floatplane wallowed at its moorings. Visibility was poor. Forest fires were burning furiously a thousand kilometers south, in Alberta, and a blue haze hung from the sky like a diaphanous curtain over the Great Slave Lake. My main purpose in traveling from Newfoundland to the town of Yellowknife on the northwestern shore of the lake in the Northwest Territories of Canada was to photograph the McDonald fault, a spectacular escarpment bordering the shoreline of the East Arm of the lake. The McDonald fault resulted from geological traumas a couple of thousand million years ago — almost half the 4,600 million years of Earth's age. These were the traumas of ancient continents in collision. Newfoundland and North Wales are doing the opposite thing:

They are separating, and the Reykjanes Ridge in Iceland had demonstrated the point. I now wanted to learn something of the structure of continents and the nature of their past movements. But before I could start taking photographs that might help to convey the drama of such geology, I needed to know more about the history of the Great Slave region from one of the scientists who specializes in this area. The Canadian Survey in Ottawa had suggested Paul F. Hoffman, who was conducting a field survey near Coronation Gulf, beyond the Arctic Circle and 600 kilometers north of Yellowknife. The large yellow and white floatplane was to deliver supplies to Hoffman's campsite that day, and Joy and I were to travel in it.

Manley Showalter was the pilot's name. He looked like my idea of the skipper of John Masefield's "Dirty British coaster with a salt-caked smoke-stack butting up the Channel in the mad March days." Instead of "Tyne coal,

road-rail . . . and cheap tin trays" the untidy, somewhat battered, and smelly interior of the cargo plane was filled with metal drums, groceries, and a handful of fishing rods, which Manley had put aboard "just in case opportunity knocks." Once in the air I stared down at the myriad lakes scattered as far as I could see, comparing what I had already learned of the Canadian Shield with the landscape below.

All continents have a nucleus of granitic rock. Granite is lighter than the basaltic materials from which the ocean crust is formed; therefore — just like freshwater icebergs in a pack of seawater ice floes — continents ride higher on the "sea" of magma beneath them and extend deeper into it than their ocean-crust counterparts do. Much has yet to be discovered about the initial formation of the continents, but scientists generally agree that it is irrevocably tied to the genesis of Earth itself.

21

In broad schematic terms the diagrams on the left and right convey the route and the objectives of the author's adventure. His goal was to discover something of the making of a continent, something of its present and past geological relationships with other continents, and something of the effect that such geological events have had on the development of life, as exemplified in North America.

The intercontinental diagram on the left cuts through the length of the British Isles, through Iceland on the Reykjanes Ridge, and through the Avalon Peninsula of Newfoundland. These places are the opening subjects of this book in the chapter entitled "Continents Apart."

Why do continents separate, and how can it be shown that they were once joined?

The title (shown in quotation marks) of each chapter in this book reflects the geological character of a particular continental province, each having a special evolutionary significance. For instance the West Coast is an "active margin" between continental and ocean crusts. Crossing the continent through the "mobile belt" of the North American Cordillera, the author's route traverses

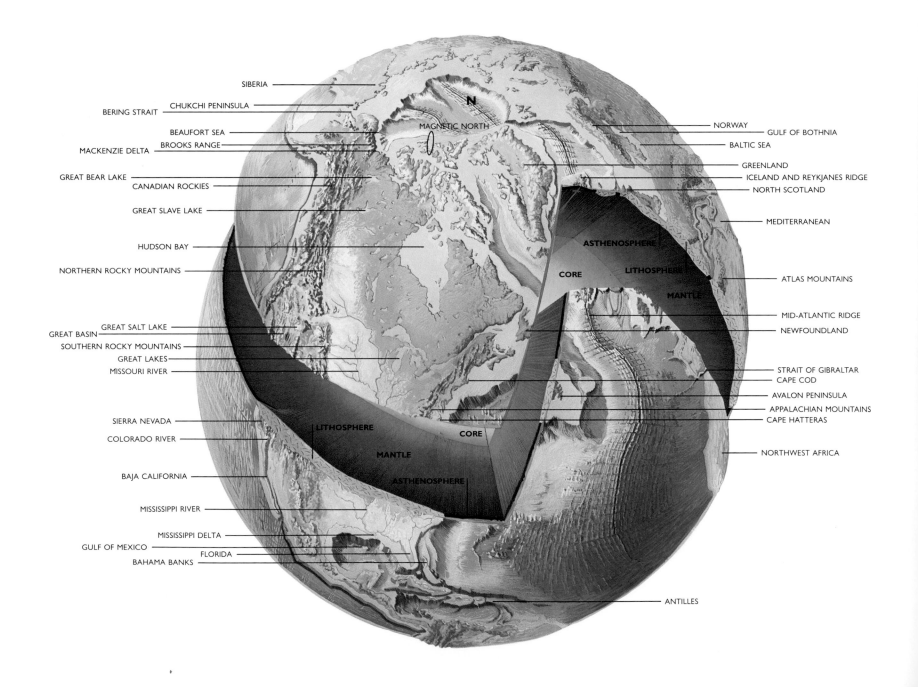

the "stable platform" and the alluvial plain through Appalachia to the "passive margin" between the eastern edge of the continent and the North Atlantic ocean crust.

The story of the continent's anatomy begins with the features ac-cented in the diagram to the right. All the geological structures of North America are either astride the nucleus of the continent, a huge feature called the Precambrian Shield, or in geological conflict with it. The shield is exposed on the surface of the continent in northern Canada, the subject of the chapter in which this illustration ap-pears. The Canadian and American Rocky Moun-tains, which stretch al-most unbroken from the Arctic Ocean to the Gulf of Mexico, form the Con-tinental Divide. As the author shows in the chapter on that subject, the effect of this range of mountains on human destiny on the continent has been profound. The diagram section ends at Yakutat on the Pacific coast of Alaska. Most of Alaska and most of the West Coast is suspect terrain because the rocks from which these regions are assembled originated elsewhere, *not* in North America. In the Pacific West from Alaska to Baja California we are seeing a continent in the making.

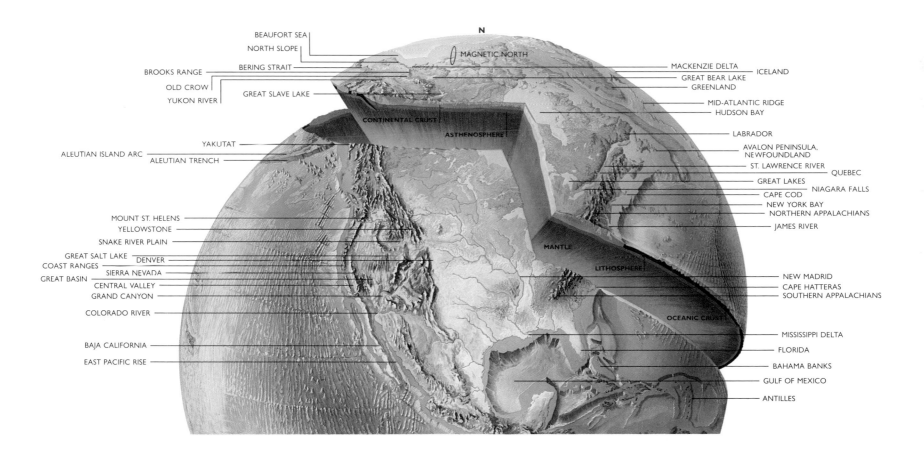

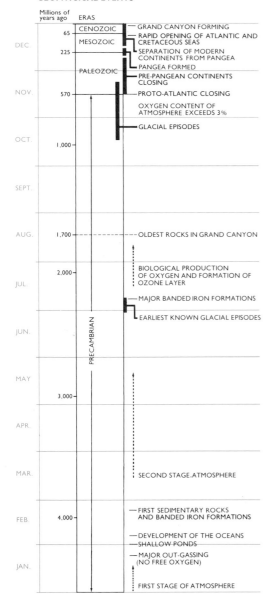

Millions of years ago	ERAS	
65	CENOZOIC	GRAND CANYON FORMING
	MESOZOIC	RAPID OPENING OF ATLANTIC AND CRETACEOUS SEAS
225		SEPARATION OF MODERN CONTINENTS FROM PANGEA
	PALEOZOIC	PANGEA FORMED
		PRE-PANGEAN CONTINENTS CLOSING
570		PROTO-ATLANTIC CLOSING
		OXYGEN CONTENT OF ATMOSPHERE EXCEEDS 3%
1,000		GLACIAL EPISODES
1,700		OLDEST ROCKS IN GRAND CANYON
2,000	PRECAMBRIAN	BIOLOGICAL PRODUCTION OF OXYGEN AND FORMATION OF OZONE LAYER
		MAJOR BANDED IRON FORMATIONS
		EARLIEST KNOWN GLACIAL EPISODES
3,000		
		SECOND STAGE.ATMOSPHERE
4,000		FIRST SEDIMENTARY ROCKS AND BANDED IRON FORMATIONS
		DEVELOPMENT OF THE OCEANS
		SHALLOW PONDS
		MAJOR OUT-GASSING (NO FREE OXYGEN)
		FIRST STAGE OF ATMOSPHERE

(left margin: ONE CALENDAR YEAR; DEC. NOV. OCT. SEPT. AUG. JUL. JUN. MAY APR. MAR. FEB. JAN.)

As interstellar matter condensed out of a rotating cloud of gas and dust with a proto-sun at its center, solid material accreted to form planetesimals, small planetary bodies, with a new, small, white star, the Sun, at the center. Collisions between planetesimals in the vicinity of the Earth's orbit ultimately produced a larger planet. The tremendous energy released by these collisions, in combination with the radioactive decomposition of some elements within the planet, eventually led to the Earth's semiliquid state. During this stage some of the heavy and most plentiful elements composing the Earth, mainly iron and nickel, accumulated at its center to form a core. As the Earth cooled by radiating heat into space, lighter elements and compounds from the still semiliquid mantle aggregated on the surface to form a thin crust, which was solid, lumpy, and comparatively cool. At this very early stage granitic material agglomerated to form the nuclei of continents, and basaltic material solidified to fill the surface between them — rather like icebergs and ice floes.

The bombardment of Earth gradually diminished as the number of smaller planetesimals, asteroids, and meteorites remaining within the Earth's gravitational influence decreased. As the Earth's crust cooled, innumerable volcanic vents opened, violently releasing volatile products trapped beneath and within the thickening crust; in turn these products formed a dense, hot, primitive atmosphere of water vapor, nitrogen, methane, ammonia, carbon dioxide, sulfur dioxide, and a host of other gases — but no oxygen in its free state. The gases condensed in the higher and therefore cooler atmosphere, producing an acidic rain that was vaporized again by the planet's hot surface. The vapor rose to a higher altitude, cooled, and reprecipitated, a cycle that contributed to the process of cooling the Earth's surface. Eventually the Earth's surface temperature fell sufficiently to permit the raindrops to form pools. As the cooling process continued, the pools became lakes, and lakes became oceans, oceans that gradually accumulated on the basalt surface between and lower than the continental masses of granite. The rain continued to fall for millions of years, reacting chemically with metallic minerals on continental surfaces to produce metallic salts in solution — the salty oceans. Solid particles of rock, some modified chemically into clays

and silts, metallic minerals, and even particles of raw metals were washed into continental lakes and onto the offshore margins between continents and ocean beds. Carbonates and metallic salts accumulated to form sediments, which were lithified into rock as one layer weighed upon another. The rugged upper surfaces of the first continental blocks were mostly eroded to low-lying surfaces ("peneplains").

From their field studies in the Harlech Dome in North Wales 19th-century geologists concluded that life on Earth began during a particular period of time that they called the "Cambrian Period," after the Roman name for Wales, Cambria. Below Cambrian rocks there were no fossils. But above them there were increasing numbers and varieties. The fossil trilobites thought at that time to be the oldest life forms in the oldest Cambrian rocks were named *Paradoxides*. Because of the easy distinction between fossil-bearing and non-fossil-bearing rocks, the expressions Cambrian and Precambrian became the first major divisions in geological time. It is now known that the oldest Cambrian Period rocks are 570 million years old and that this is disproportionately short in comparison with the now accepted 4,000 million years duration of *Precambrian* time. To handle such an enormous period of time as 4,000 million years, divisions of Precambran time became necessary. Because Precambrian continental rocks of 2,500 million years and older are distinctly different in character from their younger coun-

The Earth's upper surface is covered by lithospheric plates, consisting of heavy basaltic ocean crust and lighter granitic continental crust, which float on the plastic upper mantle of the Earth, the asthenosphere. There are seven major plates, several minor, and many microplates. Their movement on the asthenosphere is believed to be caused by complex convection-cell activity within the Earth. (See page 2.) As they move, the plates interact with each other. They rift apart, collide,

terparts, that date, 2,500 million years ago, became the next major division: The first half of Precambrian time is called "Archean," and the second half "Proterozoic." The 570 million years from the beginning of Cambrian time to the present is called "Phanerozoic."

The early Archean proto-continents were random formations of granitic lumps surrounded by basaltic seabeds. The proto-continents collided, amalgamated, and separated, perhaps more vigorously and frequently in the tumultuous circumstances of the Earth's formative years than they do in the comparative geologic calm of today. The Earth's crust during the Archean was thinner, hotter, and more unstable than it is now; it was still subject to periodic bombardment by small planetesimals and asteroids. The craters on the surface of the moon demonstrate the extent of the bombardment; for while the Earth's geological processes have continued, the Moon's surface has undergone little change since that time.

Ancient continental rocks that survived the formative period of the upper crust contain greenstone belts, huge areas of rocks dispersed among or around domelike masses of granite. They are found only in Archean rocks. Just as the appearance of trilobites signifies the end of the Proterozoic, the disappearance of the greenstone belts signifies the end of the Archean. At the beginning of the Proterozoic the characer of geological evolution was undergoing change. The interaction of the early continents with one another and with the ocean crust (in other words their *tectonic* behavior) was different during the first 2,000 million years of Earth from what it has been in the following 2,600 million years.

The nucleus of the North American continent is an amalgamation of a number of granitic protocontinents now sutured together into one huge continental block. This block is called the "Precambrian Shield," and it floats on the asthenosphere in association with the basaltic oceanic plate west of the Mid-Atlantic Ridge. The combined system is called the North American Plate, and the study of present and past interactions of this plate, other large plates, smaller plates, and microcontinental plates is the main purpose of the science of plate tectonics.

The Precambrian Shield, on which the body of the North American continent as a whole is constructed, consists of two main elements: the Greenland Shield and the Canadian Shield. Part of the Greenland Shield has been scoured by glacial ice to reveal the oldest rocks known on Earth, in an Archean greenstone belt near the present edge of its ice cap at Godthaab in western Greenland. These altered granitic rocks (Amitsoq gneisses), themselves 3,700 million years old, contain lenses of altered sedimentary rock (Isua metasediments), whose age has been estimated to be between 3,814 and 3,834 million years. The Canadian Shield was also scoured over much of its extent by glacial ice, but this ice cover has mostly disappeared, leaving much of the shield's surface free for geologists to study.

The surface of the Canadian Shield covers more than half the total area of Canada. Like the rest of the continent it is divided into geological provinces, each province having a different geological history, physiographic character, and a clearly defined boundary. The Slave Province is a good example of a geological province, for it has clearly recognizable boundary edges, consisting in part of an extraordinary wedge shape with almost perfectly straight sides. The point of the wedge and the two straight sides that make the angle penetrate the neighboring Churchill Province. The Bathurst fault marks the boundary along one straight side of Slave Province, and the McDonald fault, which I wanted to photograph, the boundary of the other. The arc that joins the two faults, the Wopmay Belt, marks the boundary between the Slave and Bear Provinces. The Slave Province is Archean in age and consists of greenstone belts and granitic rocks up to 2,500 million years old. The Churchill and Bear provinces have surface rocks that are Proterozoic, about 1,850 million years old. Why this considerable difference in age, why the unconformity? What is the reason for the wedge shape of the Slave Province? These were questions for Paul Hoffman.

Manley stuck his head around the cockpit bulkhead, the plane's engine being too noisy

subduct, or slide against each other. They behave rather like pack ice (representing the ocean crust) and icebergs (representing the continents). The active edges of the plates are called "plate boundaries" and these are illustrated in this diagram.

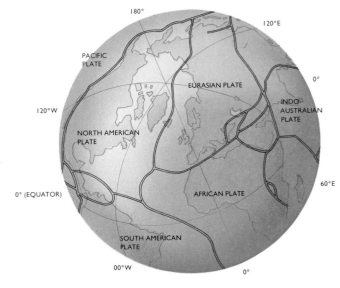

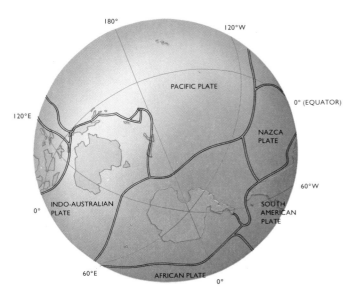

This picture illustrates a typical stretch of the surface of the Canadian Shield near Coronation Gulf above the Arctic Circle and beyond the limit of boreal forest. The CGS camp mentioned in the text can be seen in the middle distance at the left of the picture.

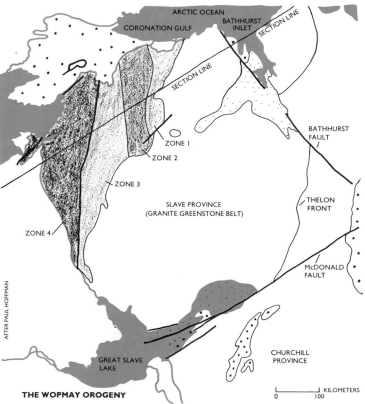

AFTER PAUL HOFFMAN

THE WOPMAY OROGENY

This sketch shows the major tectonic elements in the northwest corner of the Canadian Shield. The section line indicates the section illustrated on the opposite page.

for a human voice to carry more than an inch or two. He shouted something and pointed down to the right. I could see an arrangement of white specks in the distance. The specks were tents of the Canadian Geological Survey (CGS) decorating the flat top of a promontory of tundra jutting into a lake. Almost three hours had passed since we left Yellowknife, and we were near enough to the magnetic North Pole for a compass needle to point steeply downward. Manley had navigated the last few hundred kilometers by the seat of his pants. The Arctic air was clear, bright, and free from smoke. As we wheeled and slowed down to land on the lake, I could see Coronation Gulf on the horizon, the northernmost edge of the continental mainland and farther north from the Canadian-U.S. border than that border is from Mexico. We taxied up to a temporary jetty of planks (balanced a little doubtfully on rocks) that had been carried out from the shore. A tall, slim, athletic-looking figure waited at the safe end of the planking. Paul Hoffman held out a welcoming hand.

Mapping and geological research by Canadian Survey geologists in this region has been almost continuous for forty years. One of the pioneer participants was J. Tuzo Wilson, a leader of the avant-garde of tectonic science. Wilson's later work together with that of others had led to the conclusion that certain large tectonic events of the recent past follow a recognizable pattern. Paul Hoffman's present work in the Bear and Slave provinces suggests to him that the life cycle of an ocean basin, which he calls the "Wopmay Orogen," is also evident in this region of the Precambrian Shield, in a complete Wilson cycle. From this Hoffman concludes that the behavior of continental and oceanic plates in the Proterozoic (2,500 to 570 million years ago) was similar to that in the more recent past. With maps spread around the interior of his tent and an illustration pinned to an easel Hoffman proceeded to explain.

In a series of events that began about 2,100 million years ago and concluded approximately 1,800 million years ago, a continental plate, of which the Bear Province and the Slave Province are remnant parts, came to rest over two mantle plumes 800 kilometers apart and located in the region of what are now Coronation Gulf and Great Slave Lake. The two plumes beneath the Bear-Slave continent caused several triple-fissure-zone (triple-junction) Krafla-like events. One arm from each hot spot joined in a continuous line along what is now the Wopmay Belt. The consequence was the breakup of the original continent into the Bear continent and the Slave microcontinent (which was a sort of chip off the old block). A basaltic ocean basin developed between those two continents, a midocean spreading ridge separating them similar to the Mid-Atlantic Ridge of today. Other rifting arms emanating from the hot spots opened up troughs where Great Slave Lake now is, near the present McDonald fault, and along a line from Coronation Gulf near the present Bathurst fault.

The Great Slave Lake trough channeled sediment west toward the new Slave continental margin. As the Bear and Slave continents moved farther apart, the eastern edge of the Bear continent subsided to form a submarine continental shelf. The ocean between the two continents may have widened for 2,500 kilometers before the separation stopped. Then the continental movement reversed, causing the ocean crust between the two to subduct beneath the Bear continental plate. Eventually the Bear and the Slave continents collided, and the ocean plate that had been between them was consumed in the process of subduction. The collision between Bear and Slave caused the rigid Slave block to indent the Churchill Province lying to the southeast and to create the transform faults (one block's slipping past the other) now called the Bathurst fault and the McDonald fault.

Listening to a leading scientist explain his objectives and interpreting the facts he has gleaned during years of exhaustive exploration are always rewarding. But in this place, in a tent north of the Arctic Circle, such an experience is especially so. Paul Hoffman's work is at the leading edge of knowledge of Earth's geologic forces, in line with that of some of his predecessors in this region: John Franklin, James Clark Ross, and Roald Amundsen. These men are better known as polar explorers, but their prime objectives were scientific. They — and others of their kind — provided some of the first links in the long chain of events that resulted in the recent transforma-

tion of Wegener's controversial hypothesis of drifting continents into the hard science of plate tectonics.

It was Franklin who in 1829 reported magnetic readings taken during his exploration of the Coppermine and Mackenzie rivers. Then in 1831 Ross discovered the magnetic North Pole off the Boothia Peninsula and was surprised to notice that a needle suspended from a silk thread did not always point vertically downward over the magnetic pole: It did on some days but not on others. In 1904 Amundsen set up a long-term observatory on King William Island to study the magnetic pole which seemed to have moved geographically since Ross first discovered it. It is now known not only that the magnetic pole wanders (it is now off Bathurst Island 800 kilometers north of Ross's discovery point) but that it reverses polarity from north polar regions to south polar regions at irregular intervals — thirty times in the last 5 million years alone. Any rock with magnetically susceptible particles that has been heated during its formation or that has been reheated subsequently will take up a magnetic polarity aligned to the position of the magnetic pole at the time the heated rock cools. This phenomenon is called "paleomagnetism." Magnetic reversals recorded on each side of midocean ridges form matching paleomagnetic stripes. Comparison of these stripes proves that seabeds spread and, therefore, that continents move. Plate tectonics became established only after this discovery.

Manley Showalter was determined to fish. He made a slight detour during our return flight to Yellowknife and followed the Coppermine River upstream until he found a place he knew. We landed on the swiftly flowing river and made fast within earshot of rapids.

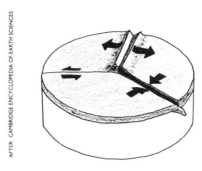

Plate boundaries are triplex, not duplex, interacting bodies. The mechanics of this interrelating movement can be visualized only at an imaginary point where three plates meet, as illustrated in the top sketch. The region above a mantle plume forms a rift ridge (top element in sketch). Two opposing surfaces form a tranverse fault (left), and two interact, as shown at the right, either by subducting or by colliding to form mountains.

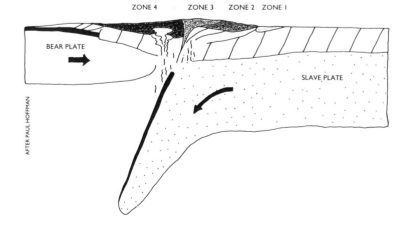

A reconstructed cross section of the continental collision between the Bear and the Slave continents during the Wopmay orogeny.

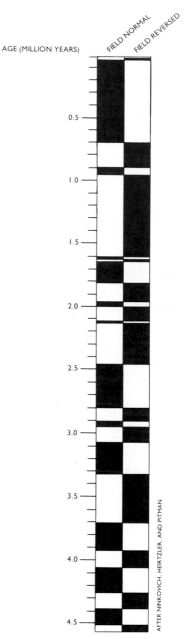

AGE (MILLION YEARS)

FIELD NORMAL

FIELD REVERSED

0.5

1.0

1.5

2.0

2.5

3.0

3.5

4.0

4.5

AFTER NINKOVICH, HEIRTZLER, AND PITMAN

We had left the "barrens," as the tundra is called in this part of the world, and were near the northernmost limit of trees. The stunted forest of the timberline was tough to penetrate and progress was slow from the floatplane to the riverbank at the foot of the rapids. Some of the willow branches were covered with what seemed to be a thick greenish moss, but when I clutched a branch for support, the moss proved to be a mass of mosquitoes. Fortunately the breeding season was at an end, and the mosquitoes did not attack; but I now believed the stories of insects so thick in the Arctic springtime that they can all but suffocate an unprotected animal or human being. Manley assembled the rods and put one into Joy's hand as she stood on a boulder. No sooner had she cast her spinner than, much to her disbelief, she hauled in a ten-pound lake trout. A second cast into the swift water, and another equally large trout! My turn: I promptly hooked a boulder on the far side and lost my spinner. But Manley, who had quietly left us to go upstream, came back in ten minutes with half a dozen fish, some of which were fifteen pounds or more. A fishy story? One of Joy's small offerings fed seven of us at dinner one evening later that week, and a good portion of that fish was still on the bone at the end of the meal.

Smoke from the forest fires in Alberta still lightly hazed the sun over Great Slave Lake when, next morning, we took off from Yellowknife in a small Cessna floatplane, our destination the McDonald fault 200 kilometers to the southeast in the East Arm of the lake. There was no wind and no horizon; the lake's placid surface simply dissolved into the cloudless gray-blue sky. We came down by Blanchet Island, one of several islands that Hoffman had marked on my map. It proved to be quite the most beautiful geologic setting I had ever seen.

The East Arm of Great Slave Lake is part of the vast channel that was formed by rifting and that thereafter directed sediments onto the Slave continental margin. Some of these sediments were precipitated from turbid water that contained a mixture of coarse and fine sands and silts eroded from the granite surface of the Slave continent. The sediments were deposited in thin beds, perhaps only a few centimeters thick, probably after very violent rainstorms. As separate depositions accumulated, they formed layers later compressed into rock, lithified by the pressure of the growing weight of overlying sediments. Such sequences are called "graywacke" formations, and although they occur throughout the geologic record, they are the most commonly found sedimentary rocks in greenstone belts.

The thin even beds of graywacke formations can be lithified only if they remain undisturbed on the seabed on which they are deposited. This suggests the kind of environment that existed when the sedimentation of the graywacke on Blanchet Island took place: very heavy rain, falling like curtain rods; fast-eroding granite surfaces and sea margins, receiving floods of heavily sedimented runoff; no waves nor currents in the sea to cause ripple marks on the seabed and no plants nor animals to disturb the even surface of the sediments; then, after the rainstorm, a period of calm perhaps like this very morning on Great Slave Lake but almost 2,000 million years ago. My delight in the beauty of the graywacke in the early morning light was heightened by awareness of the infinity of time and Paul Hoffman's account of the tectonic events in this region so long ago.

We flew low toward the McDonald fault and the Churchill Province at the eastern margin of the lake. We found it easily by following in the center of the East Arm a long curve of islands that had once been the dikes formed by the Great Slave hot spot. As the Cessna neared the high escarpment that marks the edge of the Churchill Province, the smoke haze increased. I made a number of runs over the fault, but at the right height and angle for photography the haze was too thick for the result I wanted. The smoke seemed actually to be drifting along the line of the escarpment. We landed in the bay at the foot of the escarpment and waited, but my luck was out, and visibility deteriorated: I had all the photographs I was going to get of the McDonald fault. We took off and made for another part of the East Arm, Utsingi Point at the southern end of the Pethei Peninsula on the opposite shore.

The region of Utsingi Point was another of Hoffman's suggestions; I wanted to find a good exposure of stromatolites. Over the years I had read that blue-green algae in stromatolite colonies had played a major part in changing the composition of the Earth's atmosphere during the Precambrian. At the time that the Bear–Slave continent began to rift apart, about 2,400 million years ago, the atmosphere was at a stage in its evolution when its carbon dioxide content prevailed over its oxygen content: There may in fact have been only a small proportion of free oxygen in the atmosphere at that time. But by 1,850 million years ago, when the Slave continent rammed into the Churchill Province to produce the McDonald fault, the proportion of oxygen to carbon dioxide had begun a switch that ultimately resulted in the explosive evolution of life that occurred at the end of the Precambrian, 570 million years ago.

A number of processes changed the carbon dioxide-based atmosphere into the oxygen-based atmosphere we depend upon today. But it is thought that the main factor was the prolific development of mat-like stromatolite colonies, one of the oldest forms of life on Earth, which began about 2,500 million years ago, 1,000 million years after the first blue-green algae appear in the rock record. With the aid of sunlight and photosynthesis blue-green algae could separate the hydrogen and oxygen atoms in a molecule of water to release oxygen atoms as waste material. Simultaneously they could combine the hydrogen atoms from water with carbon dioxide molecules from the atmosphere to form organic molecules — the carbohydrates, including the sugars, the starches, and the celluloses essential to animal life. Because there were no competing forms of life (no animals around to eat the algae), they became the ascendant life form. A runaway situation of algae growth developed wherever on Earth the right conditions of warmth and sunlight and adequate water occurred. That the right conditions existed here on the edge of the Slave continent was obvious by the extraordinary size of the stromatolite reefs that had accumulated off

GRAYWACKE, BLANCHET ISLAND, GREAT SLAVE LAKE, NORTHWEST TERRITORIES

These beautiful rock formations are graywacke and were photographed on Blanchet Island in the East Arm of Great Slave Lake. They were formed in the turbid waters of a shallow sea from a mixture of fine and coarse sand and from silt. They are the most commonly found sedimentary rock in greenstone belts of Archean age.

THE McDONALD FAULT, GREAT SLAVE LAKE, NORTHWEST TERRITORIES

Utsingi Point. As I walked about and climbed over these ancient limestone fossil reefs and taxied around them in the floatplane — stacks of literally millions upon millions of the mat-like stromatolite colonies that form the reefs — I understood for the first time the reality of their role in the evolution of the atmosphere and consequently upon the evolution of life.

I had planned to see two more geological features that day, a gold mine on the edge of the lake at Yellowknife and some of the best exposures of Archean pillow lavas known, to which Paul Hoffman had drawn my attention. Like the modern pillow lavas of Iceland and the Mid-Atlantic Ridge and the Cambrian pillows of North Wales and Newfoundland, the Yellowknife pillow lavas were formed by volcanic eruptions deep beneath the sea during rifting, but in an ocean 2,000 million years ago. These I had to see. And of course gold was the reason for Yellowknife's very existence.

As we cut across country to Yellowknife, I dozed in the warmth of the late afternoon sun streaming through the windshield. But when I heard "Mayday!" I sat up with sudden interest. The Cessna's single engine was sputtering. We had run out of fuel. All that maneuvering over the McDonald fault and taxiing around stromatolite reefs had caused a miscalculation, and the extra fuel we carried in cans had not been enough. The pilot made a hasty but perfectly respectable one-chance landing on Hearne Lake, one of a thousand between the East Arm and Yellowknife. We paddled to a jetty near a disused log cabin recently visited by a bear unversed in house care. It was a few hours before a familiar, large, yellow and white floatplane appeared overhead. As it landed and taxied up, the thought crossed my mind that Manley Showalter might be the pilot, and so he was. As the Cessna was being refueled, Manley's face was expressionless, apart from one eyebrow, which seemed to be permanently raised in disbelief at the cardinal sin of a bush pilot's running out of fuel. The pillow lavas and the gold mine had to wait until the following day.

"Gold!" is the magical cry that has lured

This is an ancient tranverse fault similar in character to the Great Glen fault in Scotland and the San Andreas fault in California. The main difference is one of time, not character.

The McDonald fault was active more than 2,000 million years ago when the Great Slave Plate and the Churchill Plate were moving in opposite directions during the time of the Wopmay orogeny. The surface rocks of the Churchill Province, the raised escarpment at the left of the picture, are about 1,850 million years old. The rocks of the Slave Province are up to 2,400 million years old.

These pillow lavas are over 2,000 million years old and have been worn smooth by recent glaciation. Like the modern pillow lavas of Iceland (see picture at left), they were formed beneath the sea at considerable depth; otherwise they would have exploded to form hyaloclastites — glasslike granules (see page 7, hyaloclastite beach). The pillows were formed at some time during the Wopmay orogeny and provide additional evidence that the processes that occur during plate movements today have continued through at least half of Earth's history.

STROMATOLITE REEF, GREAT SLAVE LAKE, NORTHWEST TERRITORIES

These rocks are entirely made up of fossilized stromatolite reefs of great age, on the order of 2,000 million years. Stromatolites were colonies of blue-green algae, one of the earliest forms of life on Earth. They lived at a time when the atmosphere had a carbon dioxide base. Stromatolites had no competitors and greatly multiplied wherever there was shallow saline water and warmth. They produced oxygen as a by-product of photosynthesis, by which means they also produced the carbohydrates on which they lived. As a consequence of their numbers and the amount of oxygen they gave off, they contributed largely to a change in the Earth's atmosphere to an oxygen base. This change allowed multicellular life forms to evolve and thus led to the complex series of evolutionary events that resulted in life as we know it today.

mankind to frequent catastrophe and occasional fortune across oceans and continents, especially in North America. But whatever man's attraction to it, native gold is just a mineral, one of Earth's ninety-six elements, that happens to be most frequently discovered in inconvenient and inaccessible places on the Earth's upper crust. There is a reason for the inaccessibility: The way in which gold is transferred to the near surface of the Earth ensures that it and other noble metals will be crystallized first during violent periods of crustal deformation and that chance will play the main part in its distribution and concentration thereafter.

Gold was discovered in the Yellowknife region in 1934, and the Yellowknife community, named after the Tatsanottine, a small tribe of Athabascan-speaking Indians, was established the following year. The gold-bearing quartz veins, the ore bodies in the area, had been there since the time of the Wopmay orogeny. At some time during that tectonic event an ultra-saturated water solution of silica and other minerals, including gold, had penetrated the weaknesses, the faults, and the cracks of stress and shear in the rocks as they were being folded and deformed into mountains. The most extraordinary high pressure has to exist to make the temperature of such ore-bearing water solutions high enough to dissolve their solid compounds so that, when pressure is suddenly relieved by the solutions' penetration of a rock fault, they solidify into a solid

IBYUK PINGO, NORTHWEST TERRITORIES

crystalline state and sometimes precipitate "native" elements from those compounds. In the case of the Yellowknife ore body the hydrothermal solution precipitated quartz crystals interspersed with veins of metallic gold into the overlying greenstone belt. Surface outcrops of these ore bodies were discovered to contain gold ore on the now peneplained landscape of the Slave Province. It was this discovery that led to the development of present-day gold mining operations in the Yellowknife region.

The native gold content of the upper part of the ore bodies in Slave Province was washed away by the streams that eroded the mountains formed by the Wopmay orogeny. At one stage in this process there must have been large quantities of *placer* (from the Spanish word for "shoal") gold in such mountain streams, the stuff of Klondike gold rushes. The geologic processes of 2,000 million years or so have long since dispersed such placer deposits, but there may well be deposits at the bottom of Great Slave Lake. Or perhaps gold, some ground to dust by glaciers on the Canadian Shield, may have been carried down the

ancestral drainage system of the lake, down the Mackenzie River to the Mackenzie Delta and the Beaufort Sea. Whatever the answer, millions of years from now concentrated pockets of that gold will again have been dissolved in hydrothermal solutions during a possible continental collision or buried deep in the Earth with the riverbed conglomerates in which they lie today, to become the rich gold-bearing sedimentary rocks of some future Kimberley.

On the North American continent the Mackenzie River drainage is second in size only to the Mississippi-Missouri system. Great Slave Lake is the designated source of the Mackenzie River, which runs 1,700 kilometers from its outlet at the western end of the lake at Fort Providence to the Arctic Ocean. But the lake itself receives the flow of the Peace and the Athabasca, themselves considerable rivers, which join to form the Slave River flowing into the East Arm of the lake. In addition to Great Slave Lake the Mackenzie drainage draws water from other major rivers, Great Bear Lake, and hundreds of thousands of smaller lakes scattered over nearly 2 million square kilometers. Through this network the Mackenzie basin drains the vast forested plains of northern Alberta, northeastern British Columbia, and much of the western section of the Canadian Shield. The river basin is sparse-

ly populated and has one of the most inhospitable continental climates on Earth. This is a true wilderness, perhaps the nearest we can ever know to the North America first seen by aboriginals. And that was my next pursuit: to visit what are considered by many archeologists to be the oldest human sites on the North American continent, at Old Crow Flats, on the Old Crow River in the northern Yukon. Old Crow River is a tributary of the Porcupine River, which once drained into the Mackenzie Delta but which, through a change in topography, has now become a tributary of the Yukon River.

The Mackenzie Delta is north of the Arctic Circle and is roughly triangular, with sides several hundred kilometers in length from an apex located at Point Separation. The delta's shoreline in the Beaufort Sea, the base of the inverted triangle, is about 80 kilometers long. The delta covers an area of approximately 120,000 square kilometers and is a maze of inter-twining channels, islands, isolated lakes, and swamps. A few feet below the ground the earth is permanently frozen.

Pingos and polygonal patterns, pictured here, are both manifestations of permafrost. Pingos (from the Eskimo, *pingorssarajuk*, meaning "the poor thing that is getting to be a pingo") occur most frequently in areas near the Arctic coast, like Tuktoyaktut on the edge of the Beaufort Sea. In such places ice in a depression on the surface melts in the summertime; the meltwater feeds an ice lens of permafrost, which grows beneath the surface. The lens grows in size each year until it pushes up the ground surface, the overburden, to form a hill. The largest pingos grow to more than 100 meters high. The pingo at the far left is called Ibyuk, and the basin in which it formed is estimated to be 14,000 years old.

Polygonal patterns are commonly seen from the air over permafrost country. A series of small pools of water forms on flat ground in summertime. Each pool causes a slight depression; as a consequence a larger pool forms the following year, until eventually the members of a group join their edges in the shape of polygons. Unless there is a shift in water drainage, the polygons eventually become shallow lakes, as can be seen near the edge of the picture.

There are only three communities on or near the Mackenzie Delta: Aklavik, the original settlement; Inuvik, the newly established principal community; and Tuktoyaktut at the northeastern edge of the triangle, slightly off the delta on the edge of the Beaufort Sea. The permanent pack ice of the Arctic Ocean glistened on the horizon as we turned to land at Inuvik after a flight from Yellowknife. It was late August, and the autumn colors were already fading on the aspen trees that lined the road into the colorful, modern but small town. A sudden flurry of snow caused me concern; for Old Crow was 250 kilometers farther west on the far side of the Richardson Mountains, which mark the boundary between the Mackenzie District and the Yukon Territory. August, to a southerner like me, is still midsummer, but on the delta it was already autumn, and winter loomed darkly.

The Canadian Department of Indian Affairs and Northern Development has a research facility at Inuvik, the first of three now established in the Northwest Territories. They provide field and laboratory support for scientists working in Frobisher Bay and Hudson Bay and on the Mackenzie Delta. The operations manager at Inuvik is John D. Ostrick, and when I called at the research station, he told me rather sadly that all the archeologists and anthropologists and all but one geologist, who was about to leave, had already left for home. The geologist was J. Ross Mackay, an authority on permafrost and just the man I had hoped to meet. I had been told, during a hilarious conversation one evening at the Wild Cat Cafe in Yellowknife, about the unusual things that happen in the far north. One rather well-embellished story — or so I thought — was about a permafrost manifestation called a "pingo," which has a life cycle: youth, adolescence, adulthood, and old age. Ross Mackay had written a scientific paper, "Pingos of the Tuktoyaktuk Peninsula," and fortunately Ostrick managed to arrange a meeting with him at the research center.

Permafrost is perennially frozen ground, and it can be a continuous condition, as indeed it is at Inuvik, or become progressively patchy and discontinuous, as it is farther south. Where the annual mean temperature is above the freezing point, permafrost does not exist. Here at Inuvik, where the thermometer dependably rises above freezing only during June, July, and August each year, the yearly average is minus 2°C. During the prolonged winter months Inuvik's average is minus 26°C. Two factors in particular affect the depth of permafrost: ground cover, which could be vegetation or summer meltwater, and the nature of the substrate. The coastal region of the Beaufort Sea, where a great deal of drilling has been done in recent years, has revealed permafrost to depths of 400 to 600 meters in the Tuktoyaktut Peninsula east of the Mackenzie Delta and down to 710 meters in Prudhoe Bay west of the delta. The surface above the permafrost, which thaws during summer months, is referred to as the active layer, and this can vary in depth down to 0.5 meters in regions of continuous permafrost and down to 3 meters in discontinuous zones.

Permafrost causes what southerners would view as freak effects on the landscape. But such manifestations as polygonal ice wedges are commonplace in subarctic conditions. They are elements of an Ice Age environment, and when traces of such environments are recognized in places that have a warm climate today, the discovery indicates something rather special about the past history of that region. For example other manifestations of ice ages, remnants of collapsed pingos, have been recognized in Ordovician rocks (435 to 500 million years old) in the Sahara Desert.

OLD CROW RIVER, YUKON TERRITORY

Pingos are part of the permafrost regime, but they are not commonplace because they form only in certain areas and under special conditions. *Pingo* is the Inuit word for "conical hill." *Pingorssarajuk* means "the poor thing that is getting to be a pingo," which suggests that the Eskimos have long known that pingos grow. But they grow slowly enough, under 35 centimeters a year, to make such observation difficult. The process of "getting to be a pingo" starts with the formation of a polygonal ice wedge in an active zone above the permafrost. The annual melting of the surface water in the polygon can result in the formation of a shallow lake on a bed of silt, sand, and gravel. The presence of a body of water ensures an abnormally deep active zone, which results from the combined effect of the temperature and the pressure of the water.

Erosion can cause partial drainage of the lake basin; as a consequence, in winter, a lens of ice forms below the old lake bottom in the now abnormally deep active zone. As the ice lens grows each year, the ground above the ice lens bulges upward to form a conical mound, a pingo.

Many of the larger pingos in the Tuktoyaktuk region formed in lake basins that existed 10,000 years ago. The basin in which the pingo Ibyuk (pictured on page 34) was formed is thought to be in excess of 14,000 years old. The large pingos of Tuktoyaktuk can therefore be considered sentinels from the last Ice Age. They have one common and quite distinctive feature: the composition of the mounds, their overburden. By definition the overburden must consist of materials from a residual pond, and the organic materials that originated in such ponds provide radiocarbon dates for Ibyuk and other pingos. If the water reservoir under the ice lens of a pingo is rup-

tured by erosion or by springing a leak, the pingo will collapse. Its life cycle will end. For all these reasons the discovery of a deceased pingo in Saharan rocks has a particular significance: not only that permafrost once existed in a place now desert but also that the landscape in that locality, when the pingo existed, must have been similar to the landscape on the Tuktoyaktut Peninsula today.

After a long flight in a floatplane over the British Mountains bordering the Arctic Ocean, we landed on the calm, silent, but swift-flowing water of the Old Crow River in the northern Yukon, west of the Mackenzie Delta. I clambered up the soft-silted and slippery bank to survey the site where a team of Toronto University archeologists have found what they believe to be the remains of a pingo that apparently ended its life cycle during an

The Old Crow River is a tributary of the Porcupine River in the northern Yukon Territory, north of the Arctic Circle. The Old Crow meanders over a huge area of permanently frozen ground, in which are preserved large numbers of Ice Age animal bones. With changes in course of the river, which can be clearly seen in the panorama, the bones are loosened and redeposited on river bars. Some bones that have been found by archeologists have unquestionably been worked by ancient man. The oldest authenticated finds are dated between 25,000 and 30,000 years before the present.

The smaller picture to the right shows site 0 CR 12 at the foot of the V-shaped cleft in the steep bank at the right of the picture. It is here that a University of Toronto team have discovered what they believe to be the oldest human traces yet found on the North American continent; the team claim the traces to have originated in the remains of a long-defunct pingo. The published work is very controversial because the team's dates would push back the arrival of man in North America well beyond the limits accepted as possible by other archeologists. The Toronto University date is in excess of 100,000 years before the present.

The author's handling of the bones found here was a very convincing experience. Among other things they fit the hand perfectly and have worn edges exactly where one might expect edges to be worn if the objects are held naturally. But the counterview suggests that, although some of these bones are certainly of great antiquity and may have been tools, they have been moved by the river from their original position and redeposited in an older stratum of alluvium and then covered again. They were worn into their present shape, so it is suggested, by the action of ice upon them.

interglacial period, the Sangamonian Interval, that began about 120,000 years ago and ended about 50,000 years later. Normally a pingo will die earlier rather than later in such a warm period. Even more startling than the pingo's longevity is the Toronto team's conclusion that they have also found human artifacts in the pingo debris. The artifacts consist of 223 fragments of bone. Of these, twenty fragments have features that suggest that they may have been modified by human beings, and of these twenty, three show definite signs of wear or deliberate modification. As a consequence of the finds, and many more details besides, the archeological team concluded that this locale at Old Crow was a temporary hunting campsite that predates Neanderthal Man. The hunters, possibly members of a previously unsuspected bone-age culture, probably used the pingo to give themselves a commanding view over their game on the surrounding terrain.

These are amazing conclusions, astonishing to the point that the presently accepted ideas of mankind's progress in North America would undergo a revolutionary change if the conclusions are correct. Although their position is fast being changed by other recent discoveries, most archeologists do not accept dates much beyond 12,000 years before the present for aboriginal arrivals on the continent from nearby Asia — much less dates on the order of 120,000 years. Modern man, *Homo sapiens sapiens*, dates from about 50,000 years before the present; if the Toronto University team's claims are proved, who could these denizens of Old Crow Flats have been and where did they originate? I myself had visited the Old Crow River site; at Toronto University I had handled the truly convincing bone "artifacts" that had been found there: So how is one to respond to thoroughly researched and well-presented claims that have subsequently been vigorously disavowed by other archeologists?

My panorama of the Old Crow site clearly shows how flat and wet this region is. The swift-flowing Old Crow River has to wind across its surface in a channel that silts up at every sinuous turn. Oxbow lakes and trees outline the banks of yesteryear's channel, tracing the slow but persistent changes from year to year and from century to century. In the dark days of subarctic winter this landscape is frigid. In the late days of spring the thick ice on the river breaks up and churns the soft silt on the river bottom. Old bars and banks and channels are modified. In the brief summer the river flows smoothly but strongly as the active zone above the permafrost melts and discharges surplus water. Ice scars are filled with silt; river bars are replenished; some river banks are undercut, and their soft, unconsolidated material collapses into the river. And so it has been for about 12,000 years. And between 25,000 and 30,000 years before the present this basin, like others in the northern Yukon, was partially flooded by a glacial meltwater lake. And before that, back to 120,000 years before the present, the drainage of the region was into the Mackenzie basin, not into the Yukon River, as now.

In Ottawa I consulted Richard E. Morlan, Director of the National Museum of Man, and he offered a balanced view on a complex subject. What is undisputed is that, largely because of deep-freeze preservation in permafrost substrata until a change in the course of the river channel interrupts the process, the Old Crow region presents a singular opportunity to reconstruct the plant, invertebrate, and vertebrate history of the last 100,000 years or so. But the constant reworking of the sediments by the river, the reworking of fossil bones by ice, and the transport of millions of these from one part of the basin to another where they become imbedded in permafrost again make the archeological process of sorting out and interpreting the extraordinary mixture of organic deposits on river bars and in river bluffs a challenge of the most complicated kind. The key is undoubtedly the development of accurate and reliable dating techniques. Many are being used, but none has been perfected for the difficult age ranges of the Old Crow Flats.

The dating of artifacts related to the pingo claim is based upon their stratigraphic position in the dig. The interpretation of that stratigraphy is hotly disputed. In fact, the critics say, there was no pingo: The exposed cross section is an ancient river channel, they suggest, and the few shaped and striated pieces of fossil bone claimed to be artifacts may have been altered by human beings but could equally well have been altered by the action of ice before they were redeposited in the place where they were found. The chronological limits of the consolidated layers held to be pingo remnants, the critics state, are nearer to 70,000 years before the present than to 120,000 years before the present. And a previously unrecognized bone age is thought extremely unlikely since stone artifacts have been found in association with bone artifacts elsewhere in the Old Crow basin. On the other hand, since research began in 1966, tens of thousand of fossil bones have been recovered from the region, and there is no doubt that a good many of these were shaped and used by early peoples. Recent (1982), very dependably dated fossil-bone artifacts show that human beings were here at Old Crow between 25,000 and 30,000 years before the present — still easily the earliest certainly known aboriginal people on the North American continent.

The earliest North American immigrants had to develop the technology of hunting and living in a harsh and unremitting climate. It is significant that we too have had to develop new technology to do exactly the same thing — and within a few hundred miles of Old Crow. Ancient peoples were preoccupied with the problems of killing Ice Age bison and mammoth, whereas we are preoccupied with the problems of building artificial islands in the Beaufort Sea off Tuktoyaktut to extract oil from below the seabed. In reality the hunting of bison and the drilling for oil are secondary objectives: Survival is the prime human objective, and it also seems to be the spur to technical advance.

UPPER CRUST

IT was in Dawson City, Yukon Territory, that Robert W. Service invented the character Dangerous Dan McGrew, who got himself shot in the Malamute Saloon. The scene of Service's yarn was the Klondike gold rush at the turn of the century. Today Dawson is a tourist attraction, and the frontier is 650 kilometers farther north, over the Ogilvie Mountains at Tuktoyaktut, northwest of the Mackenzie River Delta. But at Tuk one feels that the ambience of the frontier is the same today as it was at Dawson eighty years ago: intense-looking men of all sizes and shapes and colors, mostly ill kempt and unshaven, waiting for transport; half-frozen mud tracks in late August waiting for winter to turn them into passable roads; squalid, makeshift, Inuit houses intermixed with log cabins built by nouveau riche Inuit at $300 a log plus the cost of construction. The nearest trees of any size are hundreds of kilometers southward and everything has to be flown into Tuk in summer or hauled along the ice road from Inuvik in winter.

There the resemblances between Tuktoyaktut and Dawson end. Oil is the attraction now, not gold. This time the source of attraction is below the bottom of the shallow Beaufort Sea at the mouth of the Mackenzie River instead of at the bottom of Bonanza Creek off the Klondike River. The Klondike gold prospector needed only a food supply and a spade and a pan to wash gold from the stream beds — and a great deal of determined luck. The cost of taking oil from wells drilled on artificial islands in this gyrating ice-floe region of the Arctic Ocean is counted in thousands of millions of dollars. But the oil rush and the gold rush have something in common: the importance of rivers and mountains in their making.

The Mackenzie drainage system, its ancient predecessors, and its associate river systems have drained the Canadian Shield and parts of the Canadian Rocky Mountains through the ages. Between them they have removed mountains of perhaps Himalayan extent, in the form of silt, grit, and gravel from the once-mountainous shield. The alluvium was first deposited from river water to form layers of sediment on the lower exposed levels of the Precambrian basement rocks of the shield. The overflow of sediment formed deltas, similar to the Mackenzie Delta, on the margin of the continent, where continental crust met ocean crust. The enormous quantities of river-borne alluvium caused a gradual transfer of weight from the higher shield areas to lower areas, with the result that regions that accumulated sediments sank deeper into the yield-

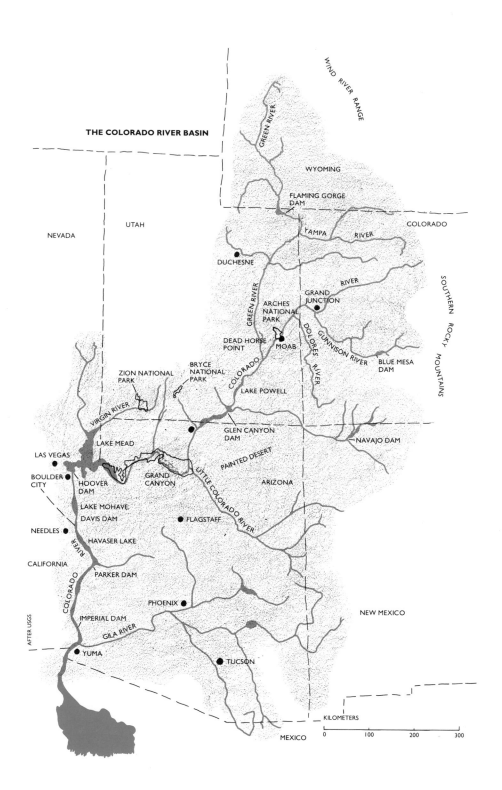

THE COLORADO RIVER BASIN

ing asthenosphere while the once-mountainous regions of the shield rebounded as they became lighter. This compensation for shifts in weight distribution on the continental crust is called "isostasy," a very short word with which to describe a very complex process: the tendency of the lithosphere to find equilibrium with the asthenosphere on which it floats.

The silts and grits and gravels that form the present Mackenzie Delta have accumulated in sedimentary layers so deep — altogether 15,000 meters or more — that they have consolidated into rock. They have also depressed the seabed with their combined weight; in a sense they have created their own basin, one far deeper than the normal level of the seabed itself. The environments in which these rocks formed often changed, and these environmental changes themselves influenced rock formation. Frequently the low-lying surfaces of the continent were inundated by shallow seas. The result was the formation of a variety of sandstone, siltstone, limestone, and shale piled one on top of the other. Organic matter constantly raining down from near the surface of shallow seas, in addition to organic materials from the land, was buried in the sediments. Once trapped, some organic materials were converted into oil and gas, "cooked" in the absence of air by the increasing pressure and heat of accumulating sedimentary layers. Ultimately the oil and gas migrated through porous rocks into reservoir rocks at the highest points in the sedimentary basin.

The formation of fossil fuels and the secretion of noble metals and other minerals were incidental. The principal result of hundreds of millions of years of periodic accumulation of sediments was the construction of a stable

SHIPROCK, NEW MEXICO

platform on the Precambrian basement. This stable platform, with a vast and slightly undulating surface, now forms the Canadian Prairies and the Great Plains adjacent to the still-exposed but considerably reduced surface of the original nucleus of the continent, the Canadian Shield. Taken as a geological unit, the basement rocks and platform rocks together are called a "craton." A cross section of the North American craton, if one were available, would reveal the geological history of the continent back to Precambrian times.

There is only one place where part of such a cross section can be seen, where by some strange trick of good fortune the Colorado River system has deeply carved the Colorado Plateau to form Grand Canyon, Zion Canyon, Bryce Canyon, and a multitude of other perhaps less well known canyons. In combination the rock sequences in these canyons now total about 4,500 meters in thickness. The river system is so deeply incised that at the bottom of Grand Canyon it has revealed the upper surface of the continent's Precambrian basement, the foundation upon which the stable platform rests.

The Colorado Plateau Province, which includes all these features, is a large, roughly round, chunk of continental craton. The rocks in this region have remained relatively undisturbed although riven with faults and surrounded by turmoil during a period when neighboring sedimentary structures were destroyed or contorted. The region is a geological anomaly. After a prolonged period of uplift it was subjected to further uplift for a period of about 5.0 million years, which began about 10.0 million years ago. During that 5.0 million years Grand Canyon and the other magnificent canyons were formed. Their rocks are *old,* but the canyons are relatively *new.*

Throughout this time the plateau was (and still is) an arid region because it is in a rain shadow caused by the high mountain range of the Sierra Nevada in California to the west. Prevailing westerly winds from the Pacific are forced high by the Sierras and lose their moisture before continuing over the deserts of Nevada to the Plateau Province. The consequent aridity severely restricts vegetation on the Colorado Plateau, the vegetation that would otherwise bind the land's surface and increase its resistance to erosion.

The Plateau Province is drained by the Colorado River system. In springtime the system is fed by the melting snows of the southern Rocky Mountains, where the Colorado originates, and from the Wind River Range in Wyoming, where the main tributary of the Colorado, the Green River, has its source. In summer sudden and heavy rainstorms unleash catastrophic flash floods. Winter cold and summer heat expand and weaken rock fissures and keep the desert surfaces friable. These conditions, operating for millions of years during the gradual elevation of the Colorado Plateau through thousands of meters, caused erosion on a scale perhaps unprecedented in

geologically recent times. The net result was established rivers' cutting deep canyons into the uplifting platform.

Why this large piece of relatively unaltered continental craton should have become separated from the main body of the craton (in the sense that it is surrounded by geological chaos) is a matter of debate among geologists. But whatever the intriguing points of discussion about the genesis of the plateau, the end result that matters most is that this natural region exists in our time for all to see. The plateau and its canyons have added immeasurably to understanding of the geological and evolutionary story of the last 570 million

The main features here in Grand Canyon are Brahma Temple (2,302 meters), Zoroaster Temple (2,173 meters), and Bright Angel Canyon, which leads down to Phantom Ranch and the Colorado River in Granite Gorge. Most visitors to Grand Canyon see these "inverted" mountains (which are carved by water out of solid rock) from the South Rim, which here forms the sharply defined horizon of this picture taken from the canyon's North Rim.

The present continents of the Earth are hybrids of the continents that preceded them. These diagrams show three of the many steps that led to the formation and separation of the supercontinent Pangea into the present six main landmasses on Earth: Africa, Antarctica, Australia, Eurasia, and North and South America.

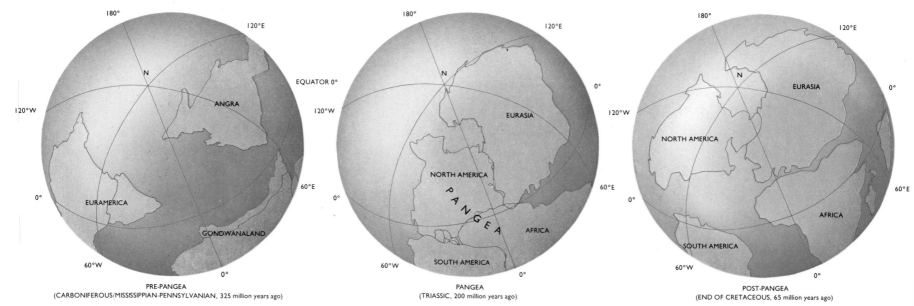

PRE-PANGEA
(CARBONIFEROUS/MISSISSIPPIAN-PENNSYLVANIAN, 325 million years ago)

PANGEA
(TRIASSIC, 200 million years ago)

POST-PANGEA
(END OF CRETACEOUS, 65 million years ago)

GRAND CANYON FROM THE NORTH RIM, ARIZONA

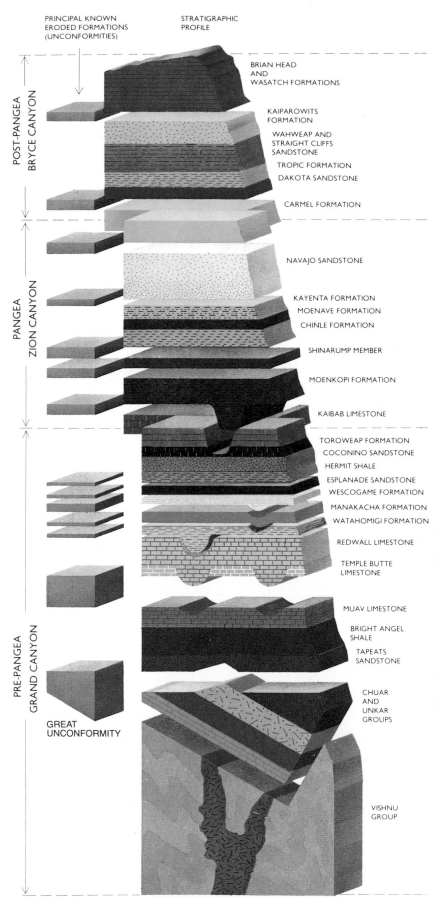

PRINCIPAL KNOWN
ERODED FORMATIONS
(UNCONFORMITIES)

STRATIGRAPHIC
PROFILE

POST-PANGEA
BRYCE CANYON

BRIAN HEAD
AND
WASATCH FORMATIONS

KAIPAROWITS
FORMATION

WAHWEAP AND
STRAIGHT CLIFFS
SANDSTONE

TROPIC FORMATION

DAKOTA SANDSTONE

CARMEL FORMATION

PANGEA
ZION CANYON

NAVAJO SANDSTONE

KAYENTA FORMATION

MOENAVE FORMATION

CHINLE FORMATION

SHINARUMP MEMBER

MOENKOPI FORMATION

KAIBAB LIMESTONE

TOROWEAP FORMATION

COCONINO SANDSTONE

HERMIT SHALE

ESPLANADE SANDSTONE

WESCOGAME FORMATION

MANAKACHA FORMATION

WATAHOMIGI FORMATION

REDWALL LIMESTONE

TEMPLE BUTTE
LIMESTONE

MUAV LIMESTONE

BRIGHT ANGEL
SHALE

TAPEATS
SANDSTONE

PRE-PANGEA
GRAND CANYON

CHUAR
AND
UNKAR
GROUPS

GREAT
UNCONFORMITY

VISHNU
GROUP

If one could cut open a section of the North American craton, one could see the ancient basement rock that underlies the whole. At the same time one could trace the development of life by examining the fossils embedded in the great number of separate rock formations that have accumulated above the basement. The rivers of the Colorado River Basin have in fact cut such a section.

The mainstream, the Colorado River, cut so deeply into the plateau that the basement rocks themselves have been revealed. This place is called Granite Gorge at the bottom of Grand Canyon. The rocks above Granite Gorge in Grand Canyon and those of Zion and Bryce canyons each represent distinct steps in the geological history of the continent. For the more technically

GRANITE GORGE, GRAND CANYON, ARIZONA

minded the stratigraphic chart to the left details the principal rock formations in the three localities. The main unconformities are also shown in this diagram: These represent periods of time, sometimes many millions of years in duration, when entire rock units were worn away.

If all the rock formations represented in the column from the Tapeats Sandstone to the Wasatch Formation could be restored one on top of the other, they would total about 5,000 meters in height. But if all the rocks that have been eroded from the plateau during the course of the last 570 million years were to be added to the pile, the total height would be more than 15,000 meters. At the bottom of the column a wedge-shaped block of rock formations is shown; this, together with the spaces surrounding it, illustrates the Great Unconformity of the Grand Canyon. The greatest gap in time represented here, between the basement rocks of the craton and the stable platform rocks, is about 1,200 million years.

PERIOD ERA

MIOCENE · CENOZOIC
EOCENE
CRETACEOUS
JURASSIC
TRIASSIC
PERMIAN
CARBONIFEROUS · PENNSYLVANIAN
MISSISSIPPIAN
SILURIAN
ORDOVICIAN
CAMBRIAN
PRECAMBRIAN

MESOZOIC DEVONIAN
PALEOZOIC CARBONIFEROUS
PROTOZOIC
ARCHEOZOIC

years, which is called the "Phanerozoic Eon." During this period life on Earth progressed from trilobite to man; many elements in this progression are recorded in the rocks of this incised section of the North American craton.

The block of limestone perched on the edge of Grand Canyon at Shoshone Point, which I mentioned in the opening paragraphs of this book, was formed in a shallow sea about 240 million years ago, when the continents of the day were on the point of completing their amalgamation into Pangea; all the rocks below it are pre-Pangean

The tiered cliffs at the eastern end of Grand Canyon are surmounted by the rocks of the Painted Desert, which began to form about 225 million years ago. To the north of Grand Canyon the same rocks that form the Painted Desert constitute the lowest tier of a series into which Zion Canyon is now cut. Broadly the rocks of the Painted Desert and Zion were deposited when North America was part of the Pangean continent.

The lowest layer of rocks underlying Bryce Canyon was formed about 130 million years ago, quite a while after Pangea split apart. The 37-million-year-old rocks of Brian Head, the highest point on the Colorado Plateau (3,400 meters), are the youngest formations remaining on the plateau today. The rocks of the Bryce Canyon area are post-Pangean.

The considerable gaps in time before and after the formation of the Pangean rocks of the Painted Desert and Zion — about 15 million years before and 50 million years after — were periods either when there was no sedimentary deposition or when formed rocks were worn away before the present rocks took their place. These time gaps between some rock formations are called "unconformities."

Pre-Pangean rocks can be seen in the deeply incised section of the plateau, Grand Canyon. These rocks rest on the Canadian Shield's Precambrian rock in the depths of Granite Gorge, cut by the Colorado River.

The Vishnu schists of Granite Gorge are about 1,700 million years old. They are Precambrian sedimentary rocks that were metamorphosed into schist during a mountain-building episode that occurred much later in time than did the Wopmay orogeny de-

scribed on page 27. But like the Slave and Bear continents which are now welded together as part of the Precambrian Shield, the original sediments that became the Vishnu schists may have formed part of a separate continent that amalgamated with others to form the Precambrian Shield. The Vishnu rocks were deeply buried, folded, and intruded with granite, Zoroaster Granite. The schist and the granite now form the steep 200-meter walls of Granite Gorge cut by the Colorado River during the last stage of major uplift. The matching near-vertical columns on either side of the gorge are believed to be the remnants of the compressed folds that once were the roots of a mountain range similar to the Swiss Alps.

The walls of Granite Gorge sometimes drop below the level of the river, disappear for long stretches between faults, and then reappear above water level. The faults cross the canyon at right angles, and this phenomenon explains the famous three sections of Grand Canyon: upper, middle, and lower Granite Gorge. The gorge itself stretches along much of the 350-kilometer length of the inner canyon. The top of it is comparatively level; the alpine mountains were reduced by erosion to a flat undulating plain (rather like the present surface of the Churchill Province next to the McDonald fault in Great Slave Lake; see page 30).

Harder rock on this surface resisted erosion when the rocks of Granite Gorge were being peneplained. Such hills of resistant rock are called "monadnocks" (after Mount Monadnock in New Hampshire), and they can still be seen at intervals along the top of Granite Gorge.

At first these were islands in a Cambrian sea, but as the sea grew deeper, the sediments accumulated on the bottom of the sea enveloped the islands entirely. The sediments formed the Tapeats Sandstone, which in Grand Canyon is the first rock of Cambrian age. Since the Vishnu schists are 1,700 million years old and the Tapeats is 570 million years old, there is an unconformity of about 1,130 million years between the Precambrian and the pre-Pangean rocks. This enormous discrepancy in age between rock layers forms part of the Great Unconformity of Grand Canyon.

Toward the eastern end of Grand Canyon, near Hance Rapids, the walls of upper Granite

Gorge disappear eastward at an angle beneath the Colorado River, but the Tapeats Sandstone continues at its original level above the river. The ever-widening wedge between the level Tapeats formation and the descending Vishnu-Zoroaster surface of upper Granite Gorge is filled with mostly red-colored sedimentary rocks set at the same angle of incline as the descending schists and granite. The sedimentary rocks are below the level of the Tapeats Sandstone and therefore are of Precambrian age. They are above the schists and granite; so they are younger than 1,700 million years. Because the reddish sediments have been intruded with igneous rock, dikes, and lavas at different times, it has been possible to bracket the age of the beds with reasonable accuracy; for they could have been intruded or covered with volcanic ash only after they had lithified into sedimentary rock. The dates produced suggest that the sediments vary in age between 600 million years and 1,100 million years, which means that there are further unconformities between these Precambrian sedimentary rocks and the Tapeats above them and the Granite Gorge rocks beneath them. These inconsistencies are another part of the Great Unconformity of Grand Canyon.

Whenever I look down into Grand Canyon, I have a vivid sense of a progression through age after age from the depths of the canyon up to the rims: the story of life's progression. From their first appearance remains of flora and fauna throughout the ages became fossilized within the sedimentary rocks of the craton. Evolution of life on Earth, as it developed on the North American craton, can therefore be traced by interpretation of the succession of such fossils in one layer of sedimentary rock above another. Geology and the study of the evolution of life are inseparable sciences, and they have been so from the time that the pioneer geologists of the 19th century first recognized what they termed the "fossil succession" in the rocks.

The oldest traces of life in Grand Canyon are found in the wedge of Precambrian sedimentary rocks where they contact the de-

The trilobite *Olenellus* is found in Grand Canyon, and *Paradoxides* is found in North Wales and Newfoundland. Such key fossils help to determine which parts of Europe and North America are geologically related.

OLENELLUS

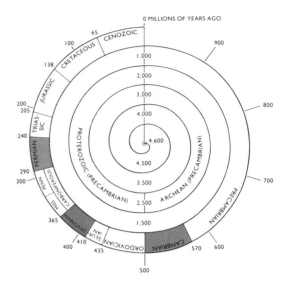

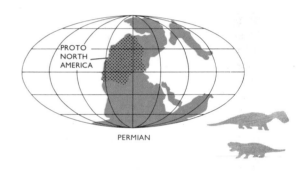

PERMIAN

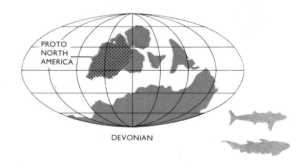

DEVONIAN

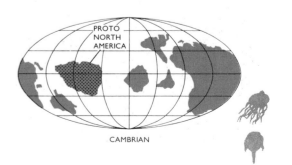

CAMBRIAN

scending rocks of upper Granite Gorge at Hance Rapids. It is in the Bass Limestone formation at Hance that remnants of a stromatolite reef have been discovered. The reef must be about 1,100 million years old, much younger than the stromatolite reefs of Great Slave Lake, which are in the order of 1,850 million years old. During the 700 million years between the lifetimes of the Great Slave and the Bass Limestone stromatolites, a vast length of time, the oxygen produced by such colonies of blue-green algae on continental margins all over Earth had continued to contribute to the conversion of the carbon dioxide–based atmosphere into one with an oxygen base. In the carbon dioxide period the variety of life forms that could develop on Earth seems to have been confined within groups of anaerobic, single-celled, prokaryotic organisms: the blue-green algae and some bacteria. But as the balance of the atmosphere changed toward an oxygen base, life began to diversify in the most extraordinary fashion. Now there developed aerobic (oxygen-dependent), multicellular, eukaryotic organisms, which could reproduce sexually. This most significant change led to the evolution of plants, invertebrates, and vertebrates.

The first Precambrian multicellular animals were creatures that evolved in saline waters, and there is little doubt that traces of them exist somewhere among the Precambrian sedimentary rocks of the Grand Canyon. But such life forms didn't have any hard parts to fossilize, and identification of amoeba-like and hydra-like fossils and the later worm-like and jellyfish-like fossils is extremely difficult because of this. But some of these early eukaryotic forms did develop hard parts, and among the first to appear in Cambrian times in Grand

Canyon are those found in the Tapeats Sandstone. They are trilobites, which swam and crawled and burrowed in the shallow-sea sediments that surrounded and eventually covered those monadnock islands on top of Granite Gorge. These trilobites, *Olenellus*, were of a genus earlier than and different from their cousins, *Paradoxides*, of the Harlech Dome in North Wales and the Avalon Peninsula of Newfoundland; but they were of the same family.

All the horizontal rock layers in Grand Canyon, from the Tapeats Sandstone to the Kaibab Limestone on the canyon's rims, were formed during the pre-Pangean period. During this time many forms of life evolved exponentially. Trilobites thrived at first, diversified, and then waned, finally becoming extinct at the end of pre-Pangean times, about 300 million years after their first appearance in the Tapeats Sandstone. Their fate was shared during the final stages of the assembly of Pangea by many other species of animals that had evolved up to that time. But surviving relatives of the trilobites developed into horseshoe crabs, sea spiders, crustaceans, and then insects; millipedes and scorpions were among the first creatures to colonize the land, and they in turn are related to the huge dragonflies that developed later, and these in turn to the primitive insects of late pre-Pangean times.

Two rock systems important to the evolution story are missing from Grand Canyon and from the plateau as a whole. Either the region was above the sea for 90 million years, or these rocks formed during that time and were later removed by erosion. The missing series are those of the Ordovician and Silurian, the periods that follow the Cambrian Period in the rock succession. Ordovician and Silurian rocks were named after Welsh tribes of Roman Britain, and there is a story to their naming: The 19th-century geologist who first identified and named Cambrian rocks, Adam Sedgwick, couldn't agree with R. I. Murchison, the geologist who first described and named the Silurian sequence, on where one series ended and the other began; after bitter dispute for thirty years it was wittily resolved that the rocks in question should be termed Ordovician since the Silures and the Ordovices had been extremely belligerent tribes in their day. But whatever the origin of their names, the fact is that the Silurian and Ordo-

MOUNT KINESAVA, ZION CANYON, UTAH

This is Mount Kinesava, one of the portals of Zion Canyon, above the town of Rockville, near the entrance of Zion National Park at Springdale, Utah. The rock formations shown here are a cross section of the mainly Pangean rocks that rise above the level of Grand Canyon's pre-Pangean rocks. The same series of rocks that form Grand Canyon's rims form the base of this mountain, and highest rocks of Zion Canyon to the northeast underlie Bryce Canyon, which is formed from post-Pangean rocks.

The main structure of Mount Kinesava is of cross-bedded Navajo Sandstone, which here forms a cliff about 750 meters thick. Before it was lithified by the pressure of other sediments upon it, the sand from which it is formed was part of a desert of vast wind-blown dunes similar in appearance to the sand seas of Saudi Arabia today.

The picture of Bryce Canyon to the right is of a part called Queen's Garden. These formations are post-Pangean lake sediments composed of shaly and easily fragmented rock pigmented by iron and other metallic oxides. The countless spires of Bryce are near the top of the sedimentary pile on the Colorado Plateau, at an elevation of nearly 3,000 meters—which largely accounts for their vulnerability to erosion.

QUEEN'S GARDEN, BRYCE CANYON, UTAH

UPPER CRUST FROM DEAD HORSE POINT, UTAH

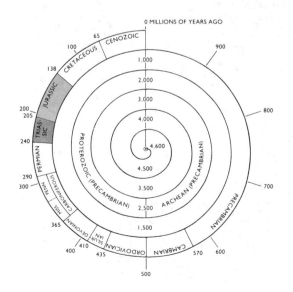

0 MILLIONS OF YEARS AGO
65
CENOZOIC
100
CRETACEOUS
900
138
1,000
JURASSIC
2,000
200
3,000
205
TRIAS-
SIC
4,000
4,600
800
240
PROTEROZOIC (PRECAMBRIAN)
4,500
PERMIAN
3,500
290
ARCHEAN (PRECAMBRIAN)
300
PENN.
CARBONIFEROUS
700
MISS.
2,500
PRECAMBRIAN
365
DEVONIAN
1,500
400
SILUR-
IAN
410
ORDOVICIAN
435
CAMBRIAN
600
500
570

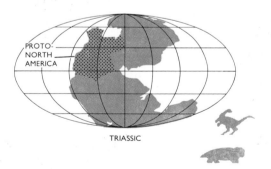

PROTO-
NORTH
AMERICA

TRIASSIC

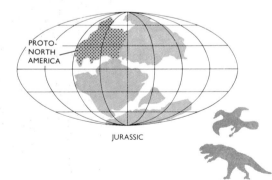

PROTO-
NORTH
AMERICA

JURASSIC

By about 240 million years ago, it is believed, all the Earth's continents had welded into a supercontinent, which remained in being for an undetermined time but upward of 100 million years. The supercontinent then split to form, first, two land masses, Laurasia and Gondwana-land, separated by the Tethys Sea, and, second, the present six great continental blocks: Eurasia, North America, Africa, Australia, Antarctica, and South America.

During much of the time in which the supercontinent Pangea existed the region of it illustrated here at Dead Horse Point in Canyonlands, Utah, was variously a humid, forested swampland, a vast sandy desert, or a shallow tropical sea or floodplain. Those ancient landscapes are today being dissected by erosion in conditions of arid desert climate similar to those which sometimes existed several hundred million years ago. The result is a scene that is the nearest we can get to seeing for ourselves what a Pangean landscape might have looked like.

In reality this scene from Dead Horse Point is one created mainly by the River Colorado's responding to the slow changes in the levels of the landscape caused, among other factors, by past movements of a sea of salt in the Paradox Basin which lies deep beneath this region. Pangean rock formations start at the White Rim in the middle distance above the highest rocks on the gooseneck promontory.

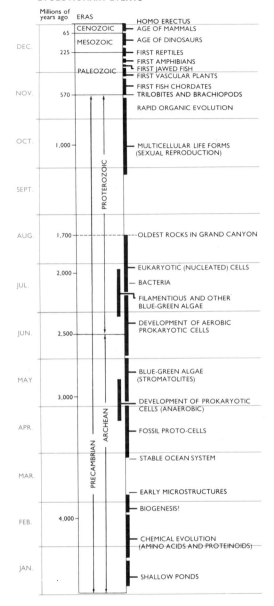

EVOLUTIONARY EVENTS

Millions of years ago — ERAS

65	CENOZOIC — HOMO ERECTUS / AGE OF MAMMALS
225	MESOZOIC — AGE OF DINOSAURS
	FIRST REPTILES
	FIRST AMPHIBIANS
	FIRST JAWED FISH
	PALEOZOIC — FIRST VASCULAR PLANTS
570	FIRST FISH CHORDATES / TRILOBITES AND BRACHIOPODS
	RAPID ORGANIC EVOLUTION
1,000	MULTICELLULAR LIFE FORMS (SEXUAL REPRODUCTION)
	PROTEROZOIC
1,700	OLDEST ROCKS IN GRAND CANYON
2,000	EUKARYOTIC (NUCLEATED) CELLS
	BACTERIA
	FILAMENTIOUS AND OTHER BLUE-GREEN ALGAE
	DEVELOPMENT OF AEROBIC PROKARYOTIC CELLS
2,500	
	BLUE-GREEN ALGAE (STROMATOLITES)
3,000	DEVELOPMENT OF PROKARYOTIC CELLS (ANAEROBIC)
	FOSSIL PROTO-CELLS
	STABLE OCEAN SYSTEM
	EARLY MICROSTRUCTURES
	BIOGENESIS?
4,000	CHEMICAL EVOLUTION (AMINO ACIDS AND PROTEINOIDS)
	SHALLOW PONDS

(ARCHEAN / PRECAMBRIAN)

Calendar months: DEC., NOV., OCT., SEPT., AUG., JUL., JUN., MAY, APR., MAR., FEB., JAN.

---

vician geological periods lasted through a crucial period of evolutionary time when simple chordates became more complex — a chain of events that after a further 400 million years of adaptation and change led to the evolution of primates and *Homo sapiens*.

A chordate is simply a living tube of eukaryotic cellular construction with a spine containing a nervous system along its length and a means of sexual reproduction somewhere near the nether end. Oxygen is assimilated through the open end at the top of the tube, which is also used for ingesting food. Digestion of food takes place during its progress through the tube, and surplus material is ejected from the end. This straightforward evolutionary invention was of such fundamental simplicity that it was almost bound to succeed. Chordates developed appendages on the outside of their tubes and a motor system, a brain, at the top of the spinal chord. In this way they became more viable in competitive aquatic environments. By the end of the Ordovician Period chordates had evolved into primitive vertebrates, the jawless fish.

During the remainder of pre-Pangean time fish developed increasingly efficient appendages: jaws, teeth, fins, and eventually legs. In some genera gill slits were replaced by lungs housed inside the body. Even so, amphibians were tied to water for breeding purposes; their eggs were blobs of jelly that could dry in the sun. Opportunity became the mother of evolution, eggs with shells developed, and amphibians became less dependent on water. Reptilian vertebrates, the first creatures to lay eggs with shells, were ready to populate the land.

One important evolutionary problem had yet to be solved. Up to this time vertebrates had evolved in large bodies of water whose temperatures changed only slowly. Reptilian body temperature was controlled by exposure to sunlight; for their surface area dictated the amount of heat that they could absorb during the day to see them through the night. An environment of hot days and cooler nights tend-

ed to favor large reptiles. But in the background—to develop strongly only toward the end of Pangean time—there began to evolve some reptilian chordates that had independent means of temperature control, which resulted in warm-bloodedness.

Animals couldn't live on land without food, and it was probably food that attracted them there in the first place. Primitive plants, like primitive vertebrates, first appeared during the Silurian Period of time, about 420 million years ago. Before their appearance it is almost certain that parts of the landscape had been colored by lichen, a life form that results from a state of domesticity between algal and fungal partners. One is prokaryotic (single-celled) and the other is eukaryotic (multi-celled), and in an extremely successful symbiotic relationship they each live on the other's by-products. Lichens have evolved that survive extremes of heat and cold, dryness and wetness, high and low altitude. Some lichens live for thousands of years, and there are at least 16,000 species on Earth today. But it is impossible to prove that they were the first living things on land because their fossil remains in sedimentary rock are difficult to identify. On the other hand land plants leave adequate fossil traces, and their record of evolution is therefore one of the most complete.

Plants evolved from the algae of the stromatolite type, which could only survive in water. The evolution of land plants required the development of eukaryotic organisms that could reproduce sexually out of water: The mosses and the liverworts had developed this capability. The next evolutionary step was the development of vascular stems consisting of tissue that could conduct water from a swamp or moist ground to a plant's extremities: The ferns and horsetails achieved this. The relatives of these early plants reproduced and grew in such enormous numbers, up to 30 meters in height and in endless variety, that they were fossilized as coal in layers sometimes tens of meters thick. One feature that characterized land plants at this time was that they were all spore reproducing, and they continued to dominate swampy regions until the end of pre-Pangean time. Although the first seed-bearing plants, the conifers, had begun to evolve, they did not really dominate the landscape until the single supercontinent of Pangea had formed. The pinyon pines on the South Rim of Grand Canyon at Shoshone Point are descendants of these first seed-bearing plants, and my block of limestone is a part of the last of Grand Canyon's pre-Pangean rock formations.

To Edwin D. McKee of the U.S. Geological Survey I owe much of my general understanding of the relationship of the evolution of life to the evolution of geology. McKee has made the canyon his own since he started working there in the late twenties and has added vastly to the geological understanding of Grand Canyon rocks and of the formation of sedimentary rocks all over the world. In addition he has found time for hundreds, maybe thousands, of people like myself who are keen students of his beloved canyon. For me Eddie McKee epitomizes great scholarship; but this gentle, unassuming man has iron resolve and stamina to match. Moreover, he has the courage and enterprise of a John Wesley Powell, the first man to traverse the Colorado River through Grand Canyon (1869).

In the 1930s, before helicopters were available, McKee and a colleague, with a party of well-known mountain climbers, formed an expedition to reach the top of Shiva Temple, near Point Sublime on the North Rim. Shiva is one of the very large (and then-unclimbed) buttes of the canyon. The expedition had been organized as a consequence of a public talk McKee had given about the separation of species that had resulted from the cutting and continued uplift of the canyon: The Kaibab squirrel of the North Rim and the Abert squirrel of the South Rim had thus evolved differently from a common ancestor. A philanthropic member of the audience had suggested and offered to finance an expedition to the unvisited top of the butte in the hope of discovering other animals that might have been similarly speciated by the separation. Exaggerated newspaper headlines excited public imagination; the parallel between the McKee expedition to Shiva and Doyle's Challenger expedition to a lost world was too good to miss. Did dinosaurs exist on the top of Shiva? If not, what strange beasts would be found? Journalists from all over America gathered on the South Rim, waiting for news to be radioed from the expedition. After "a strenuous but

an easy climb," according to McKee, animals *were* found, and they indeed proved to be anatomically different from similar species on the South Rim. The animals were *Peromiscus*: long-eared, white-tailed mice.

Once away from the rims of Grand Canyon I am much more impressed by the horizontal rather than by the vertical perspective of the plateau. The nearby deserts and canyons and particularly those which stretch for 500 kilometers northeast from Grand Canyon suggest ancient landscapes. Here, still deep in this rare section of a continental craton, the great deserts and swamps of Pangean times were once populated by reptiles, the dinosaurs. This region is a real lost world, where once there were strange-looking forests of cycads, ginkgos, and conifers, a world of vast deserts, dessicated inland seabeds, and forested swamps. Today those selfsame deserts and swamps are being eroded by wind, rainstorms, and flash floods, by the fracturing of frost, and by the natural chemical processes of erosion.

To discover a trail of three-toed dinosaur footprints in a rock, perhaps disappearing incongruously up the side of a near-vertical slab fallen from a cliff, is much more exciting than to see a reconstructed animal skeleton in a museum. It is even more impressive to visit one of the dinosaur quarry sites on the plateau. Such collections of huge, fossilized bones, often in great disarray, are sometimes the result of mass drownings of animals during a flash flood. Their bodies were swept downstream to become snagged on a river sandbar. Their bones were perhaps cleaned by scavengers, or perhaps the flesh decomposed quickly in the hot desert sun. A further flash flood disturbed the neat array of skeletons, mixed them, and buried them in river sediments, where, free from oxygen, they were fossilized by slow replacement of the organic bone with a matrix of inorganic silicate rock.

Most people are fascinated by the history of the dinosaurs, and I must count myself among those people. Edwin H. Colbert, a paleontologist who has spent a lifetime studying dinosaurs, gave me a special insight into the char-

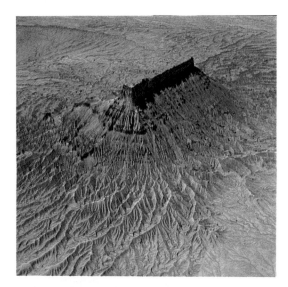

FACTORY BUTTE, ARIZONA

This is Factory Butte in the Green River Desert of Utah. It was chiseled by wind and water ero- **sion into fluted skirts. It consists largely of soft sediments laid down during the time of the earliest dinosaurs and is capped by hard rock.**

PETRIFIED FOREST, PAINTED DESERT, ARIZONA

Araucarioxylon is a species of tree of which the monkey puzzle trees, which grow naturally only in the Southern Hemisphere, are the descendants. *Araucarioxylon* forested swampy areas of Pangea. Large numbers were swept — and this is quite apparent from the aerial picture here — into this region of the Painted Desert perhaps during some catastrophic flash flood of 200 million years ago. Many of these trees were buried by silt before they could rot and were fossilized by minerals in the moist sediments in which they lay. In the course of time they were deeply buried in the colorful Chinle Formation.

At some time during their burial, after they had petrified, ground movement during earthquakes fractured them; only a few remain unbroken. When Pangea separated into southern and northern continents, Laurasia and Gondwanaland, the *Araucarioxylon* continued to evolve on the warmer southern continent. But its relatives on Laurasia could not survive the colder climates of the northerly latitudes into which Laurasia had drifted.

ARAUCARIOXYLON

acter of these creatures when I visited him in his laboratory at the Museum of Northern Arizona at Flagstaff, on the western edge of the Painted Desert. I started the conversation by raising the issue of the potential warm-bloodedness of some dinosaurs, and the answers I got to this and a host of other questions that followed was so carefully balanced a view of a subject that sometimes creates the wildest speculation that it had a profound effect on my ideas. Ned Colbert made the dinosaurs real and believable, and for that reason made them more interesting.

Alligators are modern cold-blooded reptiles. They depend on the heat of the sun for much of their energy. During the day it takes large alligators hours of basking in the sun to raise their body temperature by 1°C. At night it takes hours for them to lose that heat. Therefore the bigger an alligator, the greater the amount of heat it can accumulate and the less heat it will lose at night. Thus the larger the animal, the more even its body temperature will remain through a day and night and the more likely that it would convey the impression of warm-bloodedness even when it is not warm-blooded.

It was the same for the dinosaurs: During the first 100 million years of their existence Pangean climate and vegetation were particularly favorable to them. The bigger the animals grew, the more efficient they became in terms of the economy of their energy budget. The greater their surface area and weight, the less heat they lost through the use of their energy or a fall in temperature.

The structure of dinosaur bones, unlike that of alligator bones, is mammal-like and this could be interpreted to mean that, if not warm-blooded, dinosaurs may well have been "heterothermic" — they may have had the primitive mechanisms of blood-heat control. Indeed bone structure puts dinosaurs, among reptiles, into a class of their own. But it simply isn't known whether they were warm-blooded or not, though the chances are that some of them had this tendency.

It is the same with dinosaur intelligence. The modern crocodile is a close relative of the dinosaurs and therefore has a degree of reptilian intelligence that is at least indicative of dinosaurian intelligence. The female Nile crocodile builds a nest underground, lays her eggs, and guards the nest. When the young crocodiles begin to croak and break out of the shell, she helps them out of the nest; if one infant is having trouble, she will even take an egg in her mouth and gently crack the shell. She then takes the hatchlings in her mouth, perhaps a dozen or so at a time, and carries them down to the river to release them in the reeds, where they may be safe from predators. Dinosaurs laid eggs in nests. From the footstep imprints they left, they are known to have moved around in groups and to have displayed herd instincts. There is a parallel here, one related to instincts and intelligence.

Muscle scars on dinosaur fossil bones suggest the size and character of their muscle systems. The discovery of a large part of the skeleton of a particular species of dinosaur allows its anatomy to be reconstructed and figuratively clothed in skin. Such reconstructions are quite accurate, but they have one common failing: If a modern camel skeleton were to be fossilized, the muscle system could be worked out, but there would be no way of reconstructing the hump because it is fatty tissue and independent of bone structure. There is no way of knowing whether some dinosaurs had humps. On the other hand the fossil bones of the middle ear of dinosaurs have been recovered, and these are comparable with crocodilian stapes of today. The dinosaurs therefore had hearing, and from various inner-ear bone characteristics one can judge that some of them had acute hearing. Similar judgments can be made about eyesight and the dinosaur's ability to smell by comparison of the brain cases of known animals today with fossilized dinosaur skulls.

The dinosaurs were probably about as populous in the equatorial regions of Pangea as wild animals were populous in Africa at the end of the 19th century. The proportion of predators to prey was most likely the same, about 2 or 3 percent carnivores to herbivores; a state of dynamic equilibrium must have existed on Pangea. The herding habits were for mutual protection, and of course size was an advantage here too. A healthy adult African elephant is immune to attack from all animals except man. But even under perfect conditions for growth there is a limit to size; the weight that bones can support. As linear size doubles, the volume and weight to be supported by the legs quadruples. So the largest dinosaurs never reached the size of the largest mammals, the blue whales, simply because the weight that their bone structures could support out of water was limited by the strength of their legs. (A blue whale stranded on a beach dies from suffocation because its bone structure cannot support its weight out of the sea.) Even the dinosaurs had their limits.

The Painted Desert consists mainly of soft sedimentary layers eroded into undulating badlands. Horizontal bands of pastel reds, purples, browns, greens, and siennas permeate the graywhite base material, which is seemingly piled in indiscriminate heaps. Some Indians call the region the "land of the sleeping rainbow," which is more imaginative than the geological description, the Chinle Formation. The Chinle consists mainly of vast, forested, floodplain swamp deposits superimposed one upon the other. The trees that once forested the region, *Araucarioxylon*, are now a petrified forest of colorful agate. The animals that populated the swamps included freshwater fish, amphibians, and aquatic carnivores such as *Rutiodon*, the enormous crocodile-like dinosaur that grew up to 9 meters in length. The dinosaurs that populated this land included the three-toed, long-necked, hollow-boned *Coelophysis*, a carnivore up to 3 meters in height, which walked and ran on its hind legs, holding its prey with short front armlike legs. The upland herbivores included *Placerias*, a large heavily armored beast that fed on soft vegetation and fibrous plants.

Further northeast the Chinle gives way to a shallow sea of sandstone and siltstone and to the petrified sands of enormous Pangean deserts. The whole region, approximately 400 kilometers long and 200 kilometers wide, is stained by oxides of iron and manganese. Iron

GOOSENECKS, SAN JUAN RIVER, UTAH

The drainage from the mainly volcanic San Juan Mountains to the east of the Colorado Plateau runs into a heavily faulted area on the northern edge of the Monument Valley anticline. The ancestral river, which is now the San Juan, a tributary of the Colorado River, first formed a large lake, which was filled with sediments from the San Juan Mountains. Then during the general uplift of the Colorado Plateau the river cut down into the old lake bed and underlying rocks to form a deep meandering canyon, the Goosenecks of the San Juan River.

oxides color the rocks various shades of terracotta. Streaks of nearly black manganese oxide deposit, "desert varnish," drape the cliffs and overhangs. The total thickness and extent of these rock formations are overwhelming. For instance one formation alone, the Navajo Sandstone, was once the Pangean equivalent, in extent and depth, of the deserts of the Empty Quarter of modern Saudi Arabia. In another instance the weathering of the rock surfaces, the undulations of uplift and faulting, and the downcutting of the rivers have created a marvelous array of landforms from both Pangean and pre-Pangean rocks that have been uplifted to the surface; the latter are exemplified by Monument Valley and the Goosenecks of the San Juan River. And beyond them, back in Pangean rocks once more, the grabens, the needles, and the mazes of Canyonlands and the arches of the Entrada Sandstone.

Muley Point is an exciting place from which to view the Goosenecks and Monument Valley. The problem is that it is at the top of a mesa, a tableland with quite vertical cliffs above the sloping talus at its foot. It looked unassailable as I approached it in the motor home I used for traveling about in this country; yet the road went straight toward the mesa, and the map showed a route to the top. There simply had to be a way up. At the foot of a dirt-road incline I unhitched the jeep from the back of the vehicle so that Joy could drive it up the hill behind me. That drive in the cumbersome vehicle was quite an experience. How anyone could have imagined a road up the flanks of that mesa — let alone have built it — is beyond my understanding. Once committed, there was no turning back, and to have hesitated might have resulted in an engine stall although Joy and the jeep were there to push if need be. Foot hard down on the accelerator, in the lowest gear, tearing at hairpin bends and scraping under overhanging rock on a narrow track with precipitous edges and on a loose surface was quite Andean in character. Suddenly the road straightened out, the incline eased, and I was at the top. The view was worth the effort.

Tremendous quantities of rock have been

The region of Monument Valley was once covered to a great depth by pre-Pangean rocks similar to those to be seen below the White Rim from Dead Horse Point (pages 50 and 51). Monument Valley warped upward into an anticline. During this process rocks fractured and were eroded into mesa-like forms. As these were reduced, they became islands of rock, the familiar buttes of Monument Valley.

removed to form the familiar features of Monument Valley, which is the remaining part of an anticline. The original rocks were arched upward from beneath so that they tended to cleave. Water penetrated the clefts, and in arid climates over millions of years the whole surface of the anticline has been reduced to a series of table-topped mesas by erosion. As the sides of each mesa were worn away so that the mesa's width and length were less than its height, the mesa became a butte, or monument. After further erosion the same monuments become monoliths, which eventually topple to the desert floor. The result is one of the most spectacular and colorful of the plateau landscapes. The northern edge of the Monument Valley anticline is severely faulted; and since water always finds the lowest point and the fastest route downhill, this fault became first a lake, which filled with sediments, and then, as the plateau began to uplift, the lake bed became the site of a meandering river, which cut deeply into it to form the Goosenecks of the San Juan River.

There are pools of water among the otherwise parched rocks on Muley Point, perhaps 1,000 meters above the Goosenecks. The pools are formed from accumulated rainwater, deep enough not to evaporate immediately after a rainstorm. Later that day, while rooting among the rocks, Joy let out a shout; "Trilobites!" The creatures did look precisely as one would imagine those long-extinct creatures to look when alive. Actually, they were a remarkable subspecies of crustacean called fairy shrimp, arthropod descendants of the trilobites, with numerous jointed legs that are also gills. To swim, a fairy shrimp, in this case about 3 centimeters long, generates a wavelike movement along its series of leg gills. Its eyes are at the end of long antennae. Depending on the species, the shrimp's life cycle is completed in a few weeks, but its eggs can lie

MONUMENT VALLEY, UTAH

CYCLONE CANYON, CANYONLANDS, UTAH

Cyclone Canyon is a graben that was formed by faulting. The floor of the canyon has dropped, whereas the walls on either side have remained near their original level. There are many grabens in this locality of **Canyonlands, Utah, which have formed as a consequence of the movement of salt in the Paradox Basin deep below the surface.**

dormant for many years in desert conditions before water in a pool formed over the eggs triggers a new cycle. I later asked the experts how fairy-shrimp eggs originally got on top of Muley Point, and they suggested that such eggs can be blown great distances in the violent dust storms that occur in this region. The eggs are known to be capable of remaining dormant for fifteen years and probably much longer.

This is great jeeping country, the land of Elephant Hill and the Flint Trail. An early venture was with Kent Frost of Montecello, Utah, who knows his way around the area as well as, if not better than, anyone. Kent is a real back-country man and dresses the part. He talks in a western drawl and tells fascinating but well-embellished tales of his experiences. One of Kent's discoveries was made in a canyon (part of a system of baffling but interlinking canyons) once used by Butch Cassidy and the Wild Bunch for hiding stolen horses — or at least this is Kent's belief because Robbers Roost, a known Wild Bunch hideout, is nearby. This canyon is located in the Maze area of Canyonlands, west of the confluence of the Colorado and Green rivers. The entrance is narrow enough for a rough gate to secure horses from escape. The route from this point into the canyon system lies between circuitous narrow walls that give the Maze its name. Kent had found a direct route down the cliffs to the floor of one particular canyon.

Getting to this remote spot took a hard day's drive in Kent's four-wheel-drive vehicle. Kent had previously exploded a carefully placed dynamite charge to dispose of an overhang; so getting down the canyon walls was little more than a difficult scramble along a route marked by an occasional cairn. Once on the canyon floor, a twenty-minute walk along the sandy floor of the canyon and around a sudden bend brought us to what must be the North American Indian equivalent of the prehistoric Lascaux cave paintings. The canyon paintings are panels of pastel-colored scenes, possibly faded with time, illustrating priestly rituals. The largest of several panels is 11 meters long and 3 meters high. The face of the rock was punctured with tiny holes, which presumably were later filled with pigment, to depict tall figures that appear to be looming over an audience in front of them. Some figures have horns above dank straight hair. One has a colored striped mantle covering the

body. Some are ghostlike in the background. One figure has a tree, possibly a nut-bearing pinyon pine, growing out of an extended fingertip, with horned animals and birds nearby. Some figures are bending down as if planting corn, and in the background of the scene there is a series of straight vertical lines of varying length, perhaps illustrating standing corn.

Kent described the panel as a harvest scene, but it most probably depicts a fertility ritual of a kind still practiced by the Hopi people of today in their villages on the mesas overlooking the Painted Desert. This idea is supported not only by the subjects portrayed in the mural but also by the panel's being the background of a narrow stagelike platform at the foot of high sandstone cliffs that form a natural amphitheater. Archeologists consider it probable that Paleo-Indians of the Fremont Culture created this beautiful mural around A.D. 900. The Fremont people are thought to have moved south and to have joined with the Anasazi (literally "the old ones"), whose descendants are the modern Hopi — their village of Old Oraibi is thought to be the oldest permanently occupied human place in North America.

Directly across the Colorado River from the Maze, beneath a cliff face above a main road just north of the town of Moab, there is another colored pictograph thought to be of Fremont origin. This wall painting is locally called "All American Man" — the figure is of a rather rotund gentleman, a single feather sticking up from his head. And almost immediately opposite the pictograph there is an-

This ceremonial mural, about 11 meters long by 3 meters high, was painted by people of the Fremont Culture about 1,100 years ago, and it decorates the walls of one of the numerous canyons that form the Maze area of Canyonlands, west of the Colorado and Green river confluence in Utah.

FREMONT MURAL, CANYONLANDS, UTAH

POTASSIUM EVAPORITE PONDS, UTAH

other symbol of man, modern man. It is a uranium-ore enriching plant. Moab turned from a small desert town into a thriving community almost overnight in the 1950s when a local prospector by the name of Charlie Steen stopped to buy gasoline. The station attendant happened to have a geiger counter; it went off its scale when it was placed near some drilling cores in the back of Charlie's truck. The legend is that there was enough uranium in the cores alone to pay for the mansion that Steen subsequently built on the side of a cliff above Moab. His fabulous "Mi Vide" mine yielded many millions of dollars, which Charlie subsequently lost through unfortunate investments and tax settlements.

Northward, beyond the uranium plant, a long steep incline between high cliffs enables road traffic to surmount a doming of the local terrain. The feature that caused this uplift, which dominates the area's geology as a whole, is far beneath the surface. It is called the Paradox Basin and is the prime cause of the spectacular landforms of the region: the grabens, the needles, and the arches for which the area is renowned. To the west of the main road on the brow of the long hill there is a wedge-shaped plateau flanked by steep sand-

stone cliffs. From the top of these cliffs at Dead Horse Point, 1,800 meters above sea level and 650 meters above the Colorado River, much of the geology of the upper section of the craton formed in Pangean times has been eroded to cause an unparalleled view of the upper crust of the Earth.

In common with the rest of the Colorado Plateau the area was eroded by the interaction of the Colorado River and its tributaries and their ancestral rivers during periods of uplift. In addition to the hundreds of cubic kilometers of rock that have been removed below Dead Horse Point at least 3,000 meters thickness of mostly post-Pangean rocks has been removed from above the level of the highest plateau in sight. These volumes of rock are so vast as to be almost beyond human imagination.

About 1,500 meters beneath the region of Dead Horse Point there are 300-million-year-old rocks that correspond to some of those which were formed in Grand Canyon (see charts on pages 44 and 45). At that time the rocks now forming the Paradox Basin en-

The deep Prussian blue of the evaporating pond in this picture results from artificial colorants that are added to enhance the concentration of potassium salt. This salt originates in the Paradox Basin below the Canyonlands region and is extracted from the rocks of the Cane Creek Anticline, which can be seen above the evaporite pond at the left center of the picture.

DELICATE ARCH, ARCHES NATIONAL PARK, UTAH

The formation of an anticline in this vicinity caused the gradual fracturing of these Entrada Sandstone rocks in long jointed sections, which eroded into separate blocks and fins. These were penetrated and eroded by streams of water that percolated into the joints. As the rock was channeled into separate blocks, the wind played an increasing part in the process of erosion. The Entrada Sandstone varies in hardness and therefore erodes at different rates. Softer rock erodes first, sometimes leaving cavities in walls and fins. These grow in size, eventually to form arches capped by harder material. Pictured here is a supreme example of arch formation: Delicate Arch, in Arches National Park, Utah.

UPHEAVAL DOME, CANYONLANDS, UTAH

Upheaval Dome is the surface eruption at the top of a column of salt that reaches from the Paradox Basin far below. Such features are called salt domes, and in common with anticlines, synclines, and grabens, they are commonly associated with oil and gas exploration. Oil and gas in porous rock tend to migrate to the highest point in that rock. Although they do not form free-flowing pools and open pockets of oil and gas, the hydrocarbons accumulate in "reservoir rocks" as water does in a sponge. Such reservoirs are often near the upturned rocks that have been pushed aside by the rising column of a salt dome, at the edge of a syncline, near the top of an anticline, or beside a faulted area such as a graben. Oil exploration of these features in Canyonlands has yielded oil, but the main Paradox Basin discoveries have been on the southern margin near Four Corners. The salt domes of Texas and Louisiana and the grabens off Scotland beneath the North Sea are examples of extremely high-yield sources of this character.

closed an extensive, shallow sea, which evaporated to leave salt deposits on the floor of the basin. The basin was occasionally replenished by seawater, which also evaporated. The continuation of this process over millions of years resulted in the accumulation of very thick deposits of salt evaporites that eventually became covered with other sedimentary rocks. These rocks also accumulated to considerable thicknesses, perhaps as much as 5,000 meters or more.

Such enormous weight caused great pressure on the salt in the Paradox Basin; and pressure on crystalline salt causes it to flow like a liquid. As the weight of rock above the basin was reduced by surface erosion and faulting, the overall distribution of weight on the salt tended to unbalance; salt slowly moved about the basin beneath the overlying rocks in response to these changes. The pressures were relieved by the lifting and arching of the rock layers with the least resistance; a sort of cupola of salt collected beneath the weakest rock layers to form an anticline in the rock structure. This caused more rapid erosion of the surface rocks, which in turn increased the rate of rebound of the basement rock. To the east of Dead Horse Point one can see an exposed part of such an anticline, the Cane Creek Anticline, which has been cut by the Colorado River. In that area the 300-million-year-old sea salt is removed for industrial purposes. It is dissolved by river water and fed into evaporite ponds, where the sodium chloride (common salt) is separated from the more valuable potassium salts, as we have seen in the illustration on page 59.

If salt moved away from one area to form an anticline somewhere else, the original area might have tended either to sag into a depression, a syncline, or perhaps to fracture and fault at the surface or both. One type of structure that can result from such changes of pressure is a graben. On the surface a graben takes the form of a long, straight valley with vertical sides and a flat bottom. There are many such valleys in Canyonlands. They are normally parallel with each other, average 160 meters in width and 100 meters in height, and are often kilometers in length. Another manifestation of salt movement is a salt dome. If the salt in an evaporite basin finds a weakness in overhead rocks caused by faulting, a column of salt can intrude into that weakness and push its way to the surface. Upheaval Dome in Canyonlands is thought to be the result of such penetration.

One of the main anticlines above the Paradox Basin, Salt Valley Anticline, has the rocks of Dead Horse Point to the west of it and a large area of rock, the Entrada Sandstone, to the east of it. The Entrada Sandstone has been uplifted and gently curved on the side of the anticline. Gradual bending during uplift caused fracturing of the thick formation into thousands of elongated jointed sections. Such joints allow water to penetrate deeply beneath a rock surface that is capped by slightly harder layers of rock. In combination with the slight variations of hardness of the sandstone this has caused the area to erode into fins and blocks of rock. Some of these blocks and fins form arches as they too erode by the action of wind and water upon their exposed flanks and jointed surfaces. Which type of arch forms from a particular block or fin depends on the original size and shape of the rock, the character of the sandstone of which it is composed, and the shape and type of material with which it is capped. Some sandstone capped with a layer of harder rock results in the formation of thin, cliff-like fins, with arches beneath that have formed because of the preliminary erosion of softer components. Such arches are epitomized by Landscape Arch, with the longest known span of any arch (88 meters), and by Delicate Arch, perhaps the most beautiful natural arch of all.

The steep main road from Moab, which ascends the Salt Valley Anticline, also crosses the Green Desert to join one of the interstate highways that run from west to east. Many times this hill has heralded for me the end of a visit to the Colorado Plateau and the beginning of the long drive to Denver, Colorado, east of the Rocky Mountains. On either side of the road across the desert there are outcrops of colorful rocks, the last remaining Pangean rocks in this region. Collectively they are called the "Morrison Formation" and were named after a place called Morrison on the outskirts of Denver. The rocks here on the Colorado Plateau and in Morrison are identical in character, but a range of high mountains separates the two. The continuity was broken when Precambrian rocks, of age and type similar to the rocks of Granite Gorge at the bottom of Grand Canyon, burst through the stable platform to form the southern Rocky Mountains. The present edge of the stable platform ends at Morrison, Colorado, where it was bent upward and exposed at a steep angle when the Rocky Mountains were formed during the Laramide orogeny.

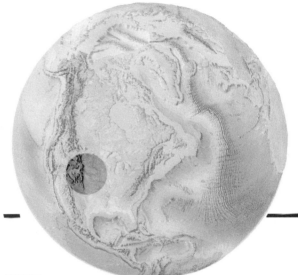

CONTINENTAL DIVIDE

FROM the Mackenzie Delta on the Arctic Ocean to the Sierra Madre Oriental along the Gulf of Mexico there is only one major break in the Rocky Mountains chain, which marks the edge of the continent's stable platform. This gap is between the Big Horn and Wind River mountains to the north and the Laramie and Front Range mountains to the south, and it separates the Middle from the Southern Rocky Mountains in the form of the wide, undulating, South Pass, 2,300 meters high. This pass over the Continental Divide descends gradually westward into the Wyoming Basin, and it was the focal point of the route followed by hundreds of thousands of pioneers in the 19th century before the transcontinental railroad was completed in 1869. If "survival of the fittest" is the catchword of evolution, it was from here to the Pacific West Coast that evolution played its most brutal

part in the great migration that populated the western third of North America.

Fort Laramie, Wyoming, at the southeastern end of South Pass acted as a valve for the stream of human beings that had crossed the 1,000 kilometers of the Great Plains from the Missouri River. Independence, St. Joseph, and Kanesville were the principal midcontinent assembly and departure points for the wagon trains that followed the trails that converged on the North Platte River, a tributary of the Missouri, westward to Laramie and into South Pass. As if they had not already had hardship enough, when these stoic and determined people reached Laramie, they shared the apprehension that from this point their real adventure began: the crossing of the Great Divide to reach their dreams of good fortune beyond.

But Laramie is more than a place-name commemorating a pioneer objective of the past and more than the name it gave to a range

of mountains. Laramie is the name geologists use to identify one of the main causes of the division of the continent by the Rocky Mountains: the Laramide orogeny. This mountain-building episode contributed to the isolation of the Colorado Plateau and the chaos around it (though not to its uplift). It caused the exposure of precious metals and other minerals manufactured in the depths of Earth, and it brought about the formation of the headwaters and the division of rivers between those that flowed east and those that flowed west. The Laramide orogeny largely determined human destiny in the Southwest. It began about 80 million years ago, and lasted about 40 million years; by its end Pangea had split up, and its elements, the modern continents, were rapidly moving toward their present geographic positions. Before the "Laramide

MOUNTAINS OF MEDICINE BOW, WYOMING

The Rocky Mountains stretch from the Arctic Ocean to the Gulf of Mexico and are geographically divided into the Canadian and the American Rockies. The American Rockies consist of the Northern, Middle, and Southern Rocky Mountains. This chapter concentrates on the Southern Rocky Mountains, which in turn consist of several mountain ranges: the Laramie Mountains and the Medicine Bow Mountains in Wyoming and the Front Range in Colorado. This picture is of Medicine Bow Peak (3,650 meters). Pioneers who named the region thought that Indians "made medicine" here and also made their bows and arrows.

Revolution," as the event is sometimes called, there were no mountains where the Rockies are today. The ancestral Rockies had been peneplained 100 million years previously, and the stable platform was a comparatively uncluttered and continuous expanse.

The small town of Morrison, Colorado, 250 kilometers due south of Laramie, is nestled in the upturned rocks of the stable platform at the foot of the Southern Rocky Mountains. Like Laramie, Morrison also lent its name to an important geological event, in this case the deposition of some of the last Pangean sedimentary rocks to be formed in this region of the North American craton. The rocks at Morrison match the rocks exposed on the surface of the Green River Desert on the western side of the Rocky Mountains; in a direct line as the crow flies the two places are about 450 kilometers apart. The same series of rocks are upturned at Dinosaur National Monument, Utah, at the edge of the Green River Desert.

Toward the end of Pangean time, from about 160 million years to 130 million years ago, the Morrison sediments now known as the Morrison Formation stretched continuously from the present town of Morrison to the Green River Desert and well beyond and around both. They partly or wholly covered areas occupied by Colorado and its neighboring states today. For millions of years the scene was one of low-lying hills interspersed with swamps and shallow lakes under a warm and dry continental climate. The red, green, gray, and tan mudstones in the Morrison Formation suggest colorful terrains that were subject to heavy rainstorms and consequently to the accumulation after flooding of thick deposits of brightly colored mud. Some of the colored mudstone that resulted from this deposition originated from windborne volcanic ash turned into mud on the ground. There was considerable volcanic activity to the west through most of Morrison time, which was also associated with the beginning of the breakup of Pangea.

Some of the largest dinosaurs ever to exist roamed about on this swampy lowland, browsing in the vegetation and wallowing in the shallow waters. Or perhaps the *Brontosaurus* and *Camarsaurus* herded together to escape predators. They and other herbivores of their time, such as *Stegasaurus*, with its thin bony

triangular plates running down the length of its spine, were hunted by one of the most vicious of the dinosaurs, *Allosaurus*. During the time that the Morrison sediments were accumulating in this region of North America there seem to have been only about a dozen or so species of dinosaur there, many fewer than were later to appear. The dinosaurs reached their zenith during the first 70 million years of post-Pangean time, the time during which Pangaea was most actively splitting up. This was the Cretaceous Period, when successive shallow seas covered the Morrison Formation and much of the continental platform. The seas deposited sandstones, limestones, and shales, which accumulated deeply to cover the Morrison sediments and further thickened the craton. During this time of episodic continental seas the dinosaurs reached their peak, after which they went into decline. Their rate of decrease exceeded their rate of reproduction, and they became extinct. In that same period the mammals began their ascent to dominance.

The first of the post-Pangean rocks that remain in the vicinity of Morrison form the rim of the hogback escarpment, which rises above the town to the east. These sandstones are covered with ripple marks on their eroded upper surface, ripples that, if it were not for their impossible angle of incline, look as though they had just been uncovered by the tide on a gently shelving beach. There are sand-filled mud cracks and the remains of sand-dune formations, which suggest periods of time above sea level. There are pebbles cemented together

(conglomerates), which were formed below the sea surface, and fossilized plant roots to mark the time of re-emergence of the seabed above the sea. There are dinosaur footprints.

The Cretaceous Period was named in 1822 by a Belgian, J. B. J. d'Omalius d'Halloy, who derived it from *creta*, the Latin word for "chalk." D'Halloy was the first to study these chalk deposits and to recognize their importance as a demarcation between two geological periods. Chalk is exposed in many parts of Europe but is nowhere better displayed than at the Seven Sisters cliffs, the famous white cliffs near Beachy Head, Sussex, on the English Channel, which for this reason is the place most strongly identified with the Cretaceous Period. The episodic continental seas that caused their formation were worldwide. *Epeiric* (from the Greek word for "continental") is the technical geological shorthand for these seas, often referred to as Cretaceous seas because of their universality in the period. All the continents were flooded within the same short periods, usually lasting 2 million years or so. Epeiric seas did not always precipitate chalk; they were equally effective producers of limestone, sandstone, and shales. Until recently the cause of such fluctuations in sea level was a very controversial subject. For example, during the 70 million years of the Cretaceous Period there were no recessional ice ages that could have influenced sea level to form epeiric seas. Although their cause is still debated, it is now broadly agreed by geologists that, because Cretaceous epeiric seas occurred during a time of relatively fast dis-

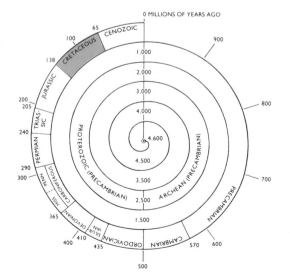

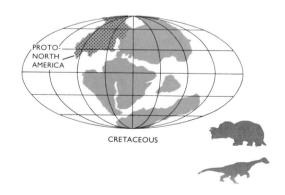

CRETACEOUS

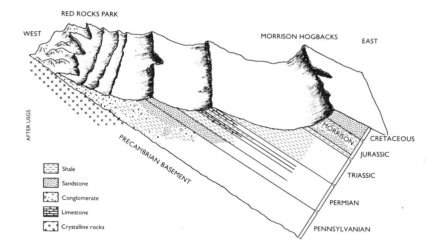

RED ROCKS PARK

WEST

MORRISON HOGBACKS

EAST

AFTER USGS

PRECAMBRIAN BASEMENT

MORRISON

CRETACEOUS

JURASSIC

TRIASSIC

PERMIAN

PENNSYLVANIAN

Shale

Sandstone

Conglomerate

Limestone

Crystalline rocks

The layers of sedimentary rock illustrated here are a cross section of those which are pictured in the Morrison hogback panorama above. They date back from post-Pangean to pre-Pangean times. The Precambrian crystalline rocks indicated at the far left represent the granites and metamorphic rocks that form the Front Range. They are basement rocks and are similar to the ancient rocks that form the lowest parts of Grand Canyon 1,000 kilometers southwest of this location.

HOGBACK ESCARPMENT, MORRISON, COLORADO

SEVEN SISTERS, SUSSEX, ENGLAND

The hogback escarpment, pictured here above the small town of Morrison, marks the edge of the piedmont between the Southern Rocky Mountains and the Great Plains. The escarpment is the upturned edge of the stable platform, once the limit of an anticline of which the Southern Rockies are the core. Morrison gave its name to the Morrison Formation, which is the last series of sedimentary rocks formed on the supercontinent of Pangea still to be found in this area; the sequence is continued on the west side of the Rockies in the Green River Desert. The rocks at the top edge of the escarpment are the first post-Pangean rocks in this region. They are ripple-marked sandstones that were formed in the Cretaceous seaways that once flooded the stable platform from the Arctic Ocean to the Gulf of Mexico.

The Great Plains stretch unbroken for 2,000 kilometers east of Denver, which can be seen toward the horizon on the right.

Cretaceous is derived from *creta*, the Latin work for "chalk." The Cretaceous Period in geological history was first recognized in Europe, where chalk formations are widespread. The best known example of chalk cliffs, illustrated here, is the Seven Sisters near Beachy Head, Susex, on the shores of the English Channel. The 70-million-year Cretaceous Period saw both the zenith of the dinosaurs, for size and numbers, and their extinction.

The red line defines the approximate limits of the area covered by the western Cretaceous Seaway, which varied in width and extent at different times through a period of 65 million years.

After the Cretaceous seas had ceased to invade the continent and after the Laramide period of mountain building was at an end, the whole of the western half of the North American continent began to uplift. Mountains, plains, and prairies—all the continental features that had already been established—were raised through thousands of meters on an axis with a center roughly coincident with the Rocky Mountains from the Arctic Ocean to the Gulf of Mexico. This section shows the uplifted regions discussed in the text.

3,850 KILOMETERS

persion of post-Pangean continents, the fluctuations in sea level may have been linked to movements of the continents. In North America epeiric seas sometimes linked the Arctic Ocean to the Gulf of Mexico. They stretched from the Mackenzie Delta to Mexico and from Utah to the Great Lakes. Most of the stable platform was immersed at different times. These seas were never stable for long, and their shorelines shifted with their advance and retreat.

This extraordinary fluctuation of sea level during an extended interglacial period is thought to have been mainly caused by changes in the capacity of the ocean basins as a result of the buildup of spreading ridges, first in the North Atlantic Ocean and later in the South Atlantic Ocean. Enormous quantities and thicknesses of basalt at the new spreading ridges simply piled up on the seafloor, displacing the water. Moreover this new basalt remained "swollen" by the heat in the region of the newly formed ridge. Thus global sea level rose during periods of great midocean-ridge activity and fell during periods of lesser activity, when the ridge regions cooled and the basalt shrank.

The extremely hot and voluminous spreading ridges may well have increased the temperature of the oceans by a few degrees. If so, this increase in temperature together with the shallowness of the epeiric seas, which naturally would be warmer than the deep ocean was,

could have had a profound effect on the sheer numbers of all forms of life and could explain the evident dramatic surge in life's evolution at that time. Recent advances in tectonic science suggest that the interrelationship between the physical world and the organic world through time has been *the* major factor in the evolutionary surge and decline of all forms of life.

The dinosaurs, for example, were victims of massive extinction at the end of the Cretaceous Period, which was itself probably the result of the splitting up of Pangea. Like many other forms of life the dinosaurs proliferated during the Cretaceous, only to become extinct at its close. Perhaps mammals, which now came to the fore, had some mechanism of survival in the new post-Pangean world that the dinosaurs did not have. In the stable climates and conditions of continental Pangea mammals had remained a minor class of life, but their characteristic adaptability assured them success in the fluctuating environments of the post-Pangean world. The consensus among the experts seems to be that the dinosaurs, unlike the mammals, had not perfected an effective system of warm-bloodedness and were therefore not so adaptable. There is probably no single explanation of the dinosaur's extinction, but it is likely to have been caused much more by a number of progessively natural events than by one single catastrophe. The extinction of dinosaurs did not happen overnight but over a prolonged period, and although some sensational possibilities have been put forward to explain their demise — supernova radiation, a major asteroid impact on Earth, the development of new viral diseases — none can account for the survival of

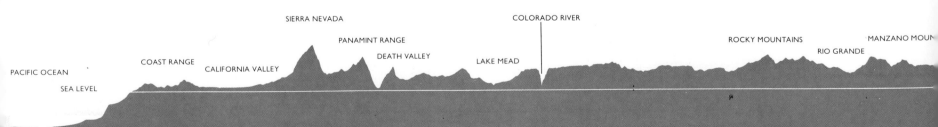

PACIFIC OCEAN

SEA LEVEL

COAST RANGE

CALIFORNIA VALLEY

SIERRA NEVADA

PANAMINT RANGE

DEATH VALLEY

LAKE MEAD

COLORADO RIVER

ROCKY MOUNTAINS

RIO GRANDE

MANZANO MOUN

other classes of life when the dinosaurs expired. Perhaps this very fact provides a clue to the most probable and most natural cause of dinosaurian extinction: the environmental changes resulting from the breakup of Pangea and the consequent climatic changes. And what more sensational event could there be than that?

The last known "species" of giant dinosaur was *Triceratops*, an evil-looking three-horned monster. A fossil specimen of this animal was found on the flanks of Green Mountain beyond the Morrison hogbacks in 1887. The local joke is that this was perhaps the very last dinosaur of all! The hogbacks to the east, Red Rocks Park to the west, and the town of Morrison in the valley between define a section through the edge of a gigantic anticline, of which the Precambrian rocks of the Southern Rocky Mountains are the core. The region of mountains to the west of the Morrison hogbacks is called the Front Range. How this range and the rest of the Southern Rocky Mountains were formed is the subject of an unresolved controversy.

There is something special about the tectonic scientists who study this kind of geological problem. Before the recent days of general acceptance of the plate-tectonic theory these men were considered the "wild bunch" of the geological profession. They deal in huge concepts. They consider the Earth's upper crust and the continents upon it as a whole. They travel widely around the planet to glean the facts and are still regarded as a little wild by the older orders of their profes-

TRICERATOPS

sion. Warren Hamilton, of the U.S. Geological Survey (USGS) in Denver, is one of these frontiersmen of tectonic science. I was lucky to catch him between trips to Jakarta, Java, and all points east.

With the aid of many maps and cartoons I gleaned from Hamilton's explanation that about 80 million years ago the Atlantic Ocean had begun to widen rapidly. Before this time North America had been moving slowly away from northwest Africa, with the result that the edge of the Pacific Ocean bed had descended at a relatively steep angle below the westward-moving continental crust, the process of subduction. But at 80 million years there seems to have been a more rapid hinge-like opening of the North Atlantic. As a consequence there was a sudden and considerable increase in the interaction between the western edge of the continent and the Pacific bed. The increase in motion and the angle of descent of the ocean crust caused part of the continental craton to rotate through a few degrees. The rotated section was the roughly circular part of the craton, which, many millions of years later, became the Colorado Plateau. The general speedup in the interaction between continental crust and ocean crust caused the ocean crust to subduct at a shallower angle than it did before. The combina-

tion of rotation of part of the craton plus the now shallow angle of subduction resulted in a tremendous upwarp of the continent. The uplift was formed where the Southern Rocky Mountains are today. These events obviously took place over an extended period of time, about 40 million years, and during this period the top surfaces of the uplift eventually eroded to form the mountains and the hogbacks we see today.

But I gathered it would be an error to relate the uplift of the Colorado Plateau itself to the formation of the Southern Rockies or to the Laramide orogeny. The plateau is related to a general uplift of the whole of western North America from Mexico to the Mackenzie Delta and from Nevada to the Mississippi. This later uplift began not earlier than 30 million years ago and included all the features that had already formed in the craton: the mountains, the plains, the prairies, and the region of the Colorado Plateau. Almost half the North American continent was simply arched upward thousands of meters on an axis roughly coincident with a line through the Rocky Mountains from end to end. Why did the whole lot uplift? There is no established answer, but Warren Hamilton's hunch is that the Pacific crust, which had been subducted at a shallow angle beneath the continent during the Laramide orogeny, was at first cold and dense. The subducted ocean crust heated up as it gradually assumed the temperature of the mantle beneath the continent. The expansion of such an enormous mass of basalt as it heated up might have caused the general uplift, "but that," Hamilton said with a twinkle in his eye, "is an extremely conjectural thesis."

LLANO ESTACADO

MISSISSIPPI BASIN

MISSISSIPPI DELTA

GULF OF MEXICO

LONG'S PEAK, COLORADO

This is the east face of Long's Peak (4,345 meters), the highest mountain in the Front Range of the Southern Rocky Mountains. It is formed from the basement rocks of the uplifted craton, from which the sedimentary rocks have been stripped by erosion. The anticline of which this is a center point splits farther north to form the Medicine Bow Mountains and the Laramie Range. The ridge of the mountain shown here forms part of the Continental Divide. Moisture precipitation on this side of the Peak flows toward the Atlantic and that from the other side of the ridge to the Pacific. The divide is an irregular line from north to south and does not necessarily follow the highest ground.

Warren Hamilton's dissertation was a bright light at the end of a rather long tunnel of technicalities. I felt that I was beginning to understand how the fundamental macromechanisms of geology determine the machinations of mankind as well as all other forms of life. The Precambrian basement rock had been squeezed to the surface with a tremendously thick covering of stable platform rocks above it. Quite likely the now-exposed basement is composed of the same character of metamorphic and igneous rocks as that which forms Granite Gorge at the bottom of Grand Canyon. That thought provoked a sudden, and for me new, awareness of the significance of the Southern Rockies. Those granite and metamorphic rocks might once have formed part of a microcontinent like the Bear, the Slave, and the Churchill provinces.

The Wopmay orogeny and the Laramide mountain-building episode, although caused by different tectonic episodes over 1,000 million years apart, had produced similar results: they had brought metallic minerals to the surface. Where the ancient Precambrian basement rocks of the Southern Rockies had faulted during mountain formation, a mineral belt had formed trending northeastward from New Mexico to Colorado. But unlike the examples of mountain formation represented by

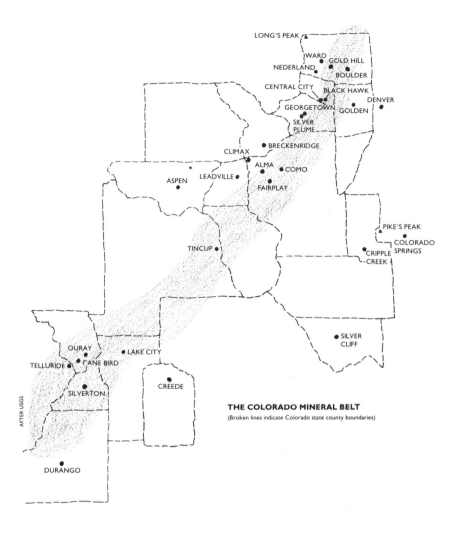

AFTER USGS

THE COLORADO MINERAL BELT
(Broken lines indicate Colorado state county boundaries)

During the Laramide orogeny the Southern Rocky Mountains were compressed and bowed upward to form an anticline. During this process the enormously thick cover of sedimentary rock was broken up and eroded to form sedimentary rocks in basins between the mountains. Also the Precambrian basement rocks of granite and metamorphic rocks were fractured and faulted. Metallic ores intruded the faults to form the mineral belt of the Southern Rocky Mountains. The belt follows the diagonal path of the fault structures.

ASPEN TREES, COLORADO

Colorado and the neighboring Rocky Mountains states are well known for the marvelous display of color provided in the autumn by aspen trees, of which this stand is typical. All mountain trees are indicators of altitude, longitude, and exposure to cold. In sheltered **places aspens in Colorado grow at an elevation between 2,400 and 2,800 meters. Near the Arctic Circle on the Mackenzie Delta they grow at sea level.**

the Wopmay event the Laramide orogeny is a comparatively recent geological phenomenon. Consequently, in the present-day Rocky Mountains it is possible to see an early stage in the history of a mountain range. The Southern Rocky Mountains sedimentary rock cover was fractured by uplift when the original anticlines were forming. It was further raised by the general continental uplift. During these two steps the original stable platform rocks had been heavily eroded by ice and snow, meltwater and streams, to fill basins that had formed between the faulted and fractured Precambrian rock core of the mountains, once the level and deeply buried shield.

Minerals from both the sediments and the shield had been washed from the rocks by those streams. For a moment in geological time gold, silver, copper, lead, zinc, tungsten, molybdenum, and other metals would be exposed on or near the surface of the mineral belt. The metallic and nonmetallic minerals were quickly reduced by erosion and dispersed. The mountains would eventually be peneplained; for that is their destiny. The perfect time for human beings to find and extract minerals from such mountains is during such a geological interval.

In 1806, Zebulon Montgomery Pike, a twenty-seven-year-old army officer, discovered Pike's Peak (4,301 meters) and reported the western extremity of the Great Plains at the foot of the Front Range to be "tracts of many leagues on which not a speck of vegetable matter existed." Fourteen years later, during an exploratory expedition, thirty-six-year-old Stephen Harriman Long of the U.S. Corps of Engineers discovered Long's Peak (4,345 meters), 160 kilometers north of Pike's Peak. Long reported this area of the Great Plains to be "wholly unfit for cultivation, and of course uninhabitable." Following visits in 1843 and 1844 on his way to and from California, John Charles Frémont made the first sensible appraisal of the mineral character of the country, but it was ignored. These not very encouraging responses to the geological challenge were crowned in the early 1850s by rumors, among others, that an Indian scout working for the U.S. Army had found gold in the headwaters of the South Platte River near Pike's Peak.

The 1850s were years of severe depression in America, and such rumors raised fond hopes. In 1858 several parties of goldminers from Georgia and about one hundred others similarly down on their luck joined forces to pan the streams that flowed east from the mountains between Pike's and Long's peaks for the rumored gold. After a month's searching without a find most of the discouraged men departed. The dozen or so who remained found placer gold on the banks of Little Creek near the junction of Cherry Creek and the South Platte River (within sight of Morrison hogbacks). They were 60 kilometers from Long's Peak and 100 kilometers from Pike's Peak. One of the men went back to "the States," meaning east of the Mississippi, boasting of his find. He displayed a small bag of gold dust, which he claimed to have found near Pike's Peak. The story found wide and exaggerated currency in the press. The gold rush that this shaky intelligence precipitated in 1858 and 1859 drew tens of thousands of prospectors, many of whom painted a slogan on their wagon sides: "Pike's Peak or bust." The emptiness of the slogan's gesture of defiance was emphasized by the fact that even Cherry Creek, the true source of the boast, had already been worked out.

At the end of the summer of 1859 perhaps

Central City and Black Hawk, pictured here, were for a time the most lucrative sources of gold and silver in the Southern Rockies mineral belt. These valleys, just below the Continental Divide at 2,550 meters, were the scene of great excitement on May 6, 1859, when John H. Gregory and Wilkes Defrees found the mother lode, the source of the placer gold that Gregory had previously found in Gregory Gulch, Black Hawk—the lower of the two towns. At the time the area was covered in aspen trees, but it was not long before rip-roaring gold-rush towns appeared together with tailings from innumerable tunnels dug into the hillsides, tailings that can still be clearly seen.

Central City, near Long's Peak, was outpaced by Cripple Creek, 160 kilometers to the south near Pike's Peak, in the 1890s, when an extremely rich find of gold and silver was made near the latter. The Cripple Creek mines are outside the mineral belt on the site of a collapsed volcano. Extraordinarily, they have yielded almost as much bullion as all the mines in the mineral belt put together.

CENTRAL CITY, COLORADO

A BEAVER DAM

GLEN CANYON DAM, UTAH

Beaver cut aspens near the ground to drag the trunks partially to block and dam a stream. Then they plug the dam from the deep-water side. Beaver gradually change the very profile and nature of a valley. Treeless and pond-covered, it becomes marshy and uninhabitable. The beaver then move on to another aspen grove.

Human beings have taken a leaf from the beaver book. The Colorado River is a much-dammed stream, and many of its dams are controversial.

If people wish to live in the southwestern desert lands of the Colorado River and to cultivate them, water is the key. The main purpose of dams is to conserve water. Electricity generated by the dams pays for their upkeep and for their original cost of construction. The prime purpose of Glen Canyon Dam (pictured here) at the end of Lake Powell is to even out the variable water supply of the great river from one year to the other—beaver-like conservation.

two or three thousand of the more determined gold seekers and entrepreneurs remained at the junction of Cherry Creek and the South Platte River on the piedmont below the Front Range. They faced the rigors of winter in tents and crude sod-insulated log cabins. Before winter set in, one man, who remains anonymous in the contemporary record, turned his oxen loose to die, for he had no winter fodder for them. To his and everyone else's astonishment the animals not only survived the winter but flourished. That burned-out-looking scrub grass was more nourishing than anyone could have guessed. With that discovery was born the stock business that was to be the mainstay of the piedmont region of the Rocky Mountains.

There were two settlements, one on each side of Cherry Creek. The settlement on the west side was Auraria, named after a gold-mining town in Georgia. The other was St. Charles until the original land claimants of the area were crowded out by William Larimer, Sr., who renamed the district Denver, after General James W. Denver, then governor of Kansas Territory, of which the whole region was then a part. There was bitter rivalry between the two communities, but for mutual benefit a reluctant unification was achieved in April, 1860, and the name of Denver prevailed over Auraria. The founding fathers knew that gold had been discovered in substantial quantities in the mountains; they had been jostling for control of what they felt to be the right place at the right time.

During the early stages of the political shenanigans between the owners of land rights on the banks of Cherry Creek, John H. Gregory from Georgia appeared in the region to take stock of local gold prospects. Unlike the hordes of "Pike's Peak Pilgrims," as the crowds of opportunists who arrived only to depart were called, Gregory was one of the few gold prospectors who really knew how to find what they were looking for. He started into the mountains in the early spring of 1859. He followed Clear Creek, a strong-flowing mountain stream that races down mountain gullies, cuts through the hogbacks, and joins the South Platte north of Cherry Creek. Placer gold had been found in Clear Creek gravel in the foothills at Golden, but Gregory was unimpressed. His objective was to discover

the source of such deposits. He followed the north fork of the swift-flowing stream high into the mountains and found rich placer-gold deposits in a gulch, now called Gregory Gulch, at Black Hawk (2,400 meters). He was lucky to survive a heavy spring snowstorm and returned to Denver for supplies and a partner. He made his way back to Black Hawk with Wilkes Defrees, another prospector, from South Bend, Indiana; together they explored the region.

On May 6, 1859, Gregory and Defrees achieved their impossible dream. They found the mother lode, the main source of the gold that had been washed down the mountain by the north fork of Clear Creek. The rich placer deposits at Black Hawk were now augmented by the discovery of the much more important gold-bearing ore in the Precambrian bedrock of the surrounding hills, the future site of Central City (2,550 meters). Within months the towns of Black Hawk and Central City were hives of industry. Clear Creek was dammed and diverted and its bed searched for nuggets and flecks of precious metal. Hundreds of holes appeared in hillsides around Central City, with tailings of yellowish rock at the entrance of each. Glory holes were blasted by dynamiting tunnels beneath irregular masses of fractured rock where the richest ore veins converged. The hillside vegetation disappeared. Smelting and reduction works were built to produce raw metal from ore-bearing quartz. A pall of smoke from the furnaces hung over the high, shallow, valleys. Saloons, gambling halls, music halls, brothels, lodgings, hotels, and a fine opera house — that extraordinary and colorful mix of squalor and splendor found in most gold-rush towns of the time — replaced the aspen groves, the tents, and the make-do shacks of the earliest days.

As the direction of the mineral belt became clearer, other finds and other mining towns sprang up with equal alacrity: Idaho Springs, Georgetown, Breckenridge, Climax, Leadville, and Aspen, to name only a few. The need for a connecting link with the outside world now became critical to their viability. Thousands of tons of low-grade ores lay ne-

DREAM LAKE, ROCKY MOUNTAIN NATIONAL PARK

Lodgepole pine decorate Dream Lake in Rocky Mountain National Park below Hallet's Peak. All the peaks on the east side of the Continental Divide were heavily glaciated during the Pleistocene Era, a period of ice ages that has lasted about 2 million years so far. We are in a glacial recession at the moment, and it is difficult to say whether this is ending or not. If the warm trend continues unabated, much of the continent will be flooded with epeiric seas; and if the glaciers return, much of the continent will be covered with ice.

SOURCE OF THE COLORADO, COLORADO

The valley in which the Colorado River rises is west of the Continental Divide beyond Central City. The North Fork of the Colorado River is the headstream. It wanders down a long, wide, sweeping valley of myriad streams and beaver-dammed lakes like the

one pictured. The photograph was taken early one October morning with ice just glazing the water's surface and with the trees on the hillside opposite reflecting an eerie yellow green in sharp contrast to the blue cold of the sheltered lake. In a few weeks the lake would be frozen, and the beaver that accounted it their home would depend on the food they had stored, plugged into the lake bottom below the ice cover and near their lodge outlet. If they had guessed wrong about the thickness of the ice that winter and it reached nearly to the bottom of their pond, the beaver wouldn't survive.

DENVER, COLORADO

glected; local industry could not economically extract the valuable metal they contained. There was no way to transport the ore across 1,000 kilometers of deserted plains to the heavy industrial plants back east.

The Union Pacific completed its transcontinental railroad in May, 1869, pushing it through Cheyenne and South Pass, Wyoming, en route to Promontory Point, near Salt Lake in Utah, and the West Coast. This route through the mountains roughly followed the route of the wagon trains that the railroad was intended to replace. The line was a long way north of Denver and the Front Range communities. The founding fathers of Denver were incredulous. As the head of the new railroad had gradually progressed across the plains, no amount of persuasion had succeeded in getting the Union Pacific routed through Denver. The city was simply too far south. It was in the wrong place — all those horrible mountains between it and the West Coast. The frustrated Coloradans financed and built their own connection from Denver

to Cheyenne (the Denver Pacific Railroad), but it went bankrupt shortly after its completion in 1870. After extended legal squabbling the Union Pacific constructed a parallel line from Cheyenne to Denver in 1877.

During this period Denver and Golden were vying for supremacy; one community or the other one day had to become the state capital. Golden had the political support of the mining towns: It understood mining problems and was also on Clear Creek at the entrance to the mountain highway now constructed beside the stream. Golden was literally the gateway to the mountains. But Denver's founders had set out to attract first-class lawyers to their growing community. They had had the foresight to anticipate that litigation would be one of the penalties of unread mining titles; lawyers would therefore play an essential part in Denver's future as the service center and capital of the region. Conversion of gold to money and property and the need to develop mining-related industry had attracted bankers, real-estate men, and speculators as well as engineers and draftsmen and designers. The importance of news from the mining communities led to the founding

of newspapers on the banks of Cherry Creek, of which the *Rocky Mountain News* (1859) is the only survivor. The result of the bitter struggle was that Golden lost the commercial battle but won the intellectual one: It became the site of the Colorado School of Mines and the USGS. Denver became the capital of the territory in 1867 and of the state of Colorado in 1876.

Pike's Peak eventually came up with the richest gold find of all. Two shepherds picked up richly mineralized rocks near Cripple Creek high in the hills to the southwest of the mountain. The gold mines that eventually peppered the region produced almost as much gold and silver (gold and silver are often found together, and this was generally true in Colorado) as all the other sources in the Southern Rocky Mountains mineral belt put together. But the Cripple Creek discovery was made thirty years after the Pike's-Peak-or-bust episode, and even then the discovery was in an anomalous region, an area outside the mineral belt. The Cripple Creek lodes

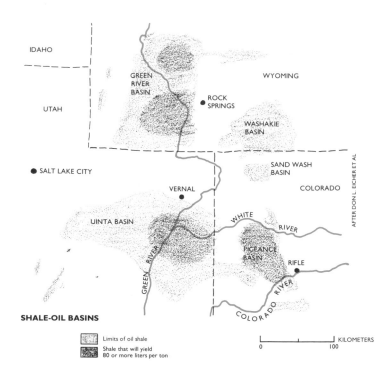

SHALE-OIL BASINS

Limits of oil shale

Shale that will yield
80 or more liters per ton

KILOMETERS
0   100

AFTER DON L. EICHER ET AL

Denver, Colorado, is widely known as the Mile High City. Because of its altitude and its being in a rain shadow caused by the Front Range, it has a remarkably dry climate, which makes tolerable both winter cold and summer heat. In the hundred years or so since Denver's founding on the banks of Cherry Creek, the city has been the center of one mineral boom after another, all thanks to the natural resources of the southern Rockies. Today the gold and silver mines are more or less worked out. The present boom is concerned with oil and gas exploration and recovery. The biggest prize and the most difficult to achieve is the extraction of oil from lake shales formed when the Laramide orogeny was nearing its close. The reserves of shale oil are estimated to be larger than all the oil ever to have been humanly used.

The new buildings to house the large companies that the oil boom has attracted to the region have transformed the appearance of the city center. Most of the high buildings in the picture did not exist before 1978.

were formed around 40 million years ago at the site of a collapsed volcano, a caldera, remains of one of the last flings of the Laramide orogeny before it subsided into grumbling quiescence.

When the Denver and Rio Grande Western Railway, built with a view to establishing an alternative to the Union Pacific route to the West, reached the eastern base of Pike's Peak in 1871, just one dwelling existed in the locality. It was a low, flat, mud-roofed log-cabin hotel owned by Richard Sopris, a future mayor of Denver. This hotel became the nucleus of the city of Colorado Springs, an elegant substitution for the original name of Manitou Mineral Springs. The founder of the new town was William J. Palmer, who was building the railroad. He and his associates set out to establish a community that would be an intellectual and social cut above other burgeoning places in Colorado. With a sound sense of good public relations the infant town was dubbed "the Athens of Colorado." By 1880 a flourishing city of almost 10,000 people had been established. But although this was fast growth by any standard, it was minuscule in comparison with the population growth on the piedmont as a whole. By 1890 some 150,000 people lived on the narrow strip on the edge of the stable platform between Pike's Peak and Long's Peak; within ninety years the number had increased twentyfold. There was not enough water to supply the needs of such a population.

The western slope beyond the Continental Divide has many broad and well-rounded valleys, whereas the eastern slope has valleys that are generally steep and narrow. The west slope not only collects several times as much moisture as does the east slope but also collects more heat from the sun. The result is a faster rate of erosion on the west slope, which accounts for the different character of the landscapes on opposite sides of the Continental Divide. The headwaters of the Colorado River, west of Long's Peak, are tantalizingly close to the Colorado piedmont, whose eastern slopes thirst for sufficient water to meet the needs of land irrigation for crops grown in a semi-arid climate and a growing population. Without more water than could be supplied by groundwater aquifers and artesian wells on the eastern side of the Front Range, plus all the reservoir storage capacity possible east of the divide, the piedmont would fast become a desert.

The solution was to build tunnels from west to east beneath the mountains. One of these is 24 kilometers in length. It draws water from Grand Lake near the source of the Colorado River and carries the water almost beneath Long's Peak to the eastern side of the Front Range at Estes Park, the town associated with Rocky Mountain National Park. Another scheme transfers water from Lake Dillon, under the very mountains that once yielded most of the mineral wealth of the state of Colorado, this time through a tunnel 35 kilometers long into the north fork of the South Platte River. But ultimately there has to be a limit to the water drawn from west of the watershed to the east slope. Consideration has to be shown for the needs of many other western regions that are absolutely dependent on Colorado River water. Water supply in the arid American Southwest is political dynamite.

An inescapable problem that faces not just the Southwest but North America and most other continents besides is that of recovering from the Earth the means of producing adequate energy for man's continued existence. At least part of the answer to this problem lies buried in lake-bottom shales formed during the Laramide revolution — perhaps that event's parting gesture. At Bryce Canyon, Utah, rocks are exposed that are near the top of the sedimentary pile on the Colorado Plateau (see page 44). Bryce Canyon is a vividly colorful product of erosion, carved from lake-bottom shales formed during the same period of time, in similar lakes, and in the same general region of the southern Rockies as were the shales of the Green River and Piceance basins, which are estimated to contain more oil than has ever been used since the invention of its use — in excess of 4 million million barrels. The objective of some of the men and women in the glistening new glass, steel, and concrete towers in Denver is to get the oil out of that shale although a temporary glut of oil from other sources has delayed development.

The new pioneers arrived in jet airliners instead of horse-drawn wagons. They probed the geological history of the Paradox, the Big Horn, the Wyoming, and the Denver basins to plan the exploration and the extraction (if they are lucky) of conventional oil and gas as well as to ponder the problems of shale oil, which are immense. The gold and silver has mostly gone. Zinc, lead, tungsten, molybdenum, and uranium are out of fashion. Oil is in. Fort Laramie is a museum, the trail over South Pass a curiosity, and the railroads in decline. But yet another mineral boom has hit Denver by Cherry Creek — at least for the moment.

SUSPECT TERRAIN

THE large man smiled through a forest of rolled maps tucked in the crook of one arm as he struggled to maintain a tenuous grip on a voluminous folder clutched under the other. His free hand was extended in my direction. "Hi! My name is Gil Mull." Guardian angels come in all sizes.

Here in a corner suite in the upper regions of the Captain Cook Hotel overlooking Cook Inlet in Alaska, one of the geologists who had been a main contributor to the exploration for, and discovery of, oil on the North Slope had arrived to perform a mission of mercy. My plans had gone sadly awry. The hotel computer had somehow mangled my reservations, and my room, the last available during a convention week, had been let to someone else. The suite, at no extra charge, plus a basket of fruit and a couple of bottles of wine

with a note of apology were the management's way of compensation. Impressive! My intention had been to drive up the Haul Road from Fairbanks through the Brooks Range to the North Slope; it had been confounded. The Sohio Alaska Petroleum Company's operations manager, Roger Herrera, who had so generously undertaken to obtain a permit, was stunned by the state bureaucracy's refusal to allow me to drive the admittedly difficult road, which follows the Alaskan Pipeline along 650 kilometers from central Alaska to the shores of the Arctic Ocean. Herrera offered to fly Joy and me up to the North Slope from Anchorage and to provide an introduction to C. G. Mull, formerly of the U.S. Geological Survey (USGS) and now a bastion of the Alaska Geological Survey. Brilliant! The Captain Cook Hotel suite was a delight; the North Slope visit was assured; and Gil Mull was the answer to a prayer.

What I hadn't appreciated before I met him

was that Gil is a widely published photographer as well as a geologist: He understood my need to grasp the broad geological concepts first and then to discuss photographic objectives. His help was indispensable; for Alaska is immense, seven times the area of Great Britain and larger than California, Arizona, New Mexico, Utah, and Colorado combined. Almost at once we were on our hands and knees among every kind of map, cartoon, and print.

I first gathered that, although the Brooks Range and the North Slope were once part of the North American craton, the rest of Alaska is mostly made up of bits and pieces, microcontinents, and ocean crust. A sort of Alice in Wonderland feeling overtook me, a sense that the more I learned the less I knew. It seems that, around the time that rifts began to form — the prelude to the breakup of Pangea about 140 million years ago — the North American craton extended considerably northward beyond the Mackenzie Delta. The continental basement and stable platform projected at least 1,000 kilometers toward the North Pole. A ridge in the Canada Basin beneath the Arctic Ocean, rather like the Mid-Atlantic Ridge today, began to spread beneath this part of the craton, causing a wedge-shaped gap to open in it. The opening was wide toward the Pole and narrow toward the Mackenzie Delta. The delta region was the axis of the hinge. As the left arm of the rifted region rotated westward its surface was frequently submerged beneath shallow Cretaceous seas, from which limestones, sandstones, and shales were deposited in a way similar to, and during the same period as, those of the Colorado Plateau and the Southern Rocky Mountains provinces to be (the Laramide orogeny had not yet begun). But unlike their midcontinent counterparts the sedimentary rocks laid down on the rotating arm overlapped at their extremities like a pile of neatly stacked playing cards carelessly pushed to one side. These sediments were deposited at a time of superabundant life. The shorelines of the western Cretaceous Seaway (see page 68) were thick with shallow-marine micro-faunal algae — literally, an organic soup of incredible richness.

As the piece of craton that later formed the Brooks Range and the North Slope was completing its swing, it began to interact with the Pacific Ocean Plate. The ocean crust simply overrode the continental crust on the continent's now-south-facing margin, an interaction some geologists call "obduction." The continental sedimentary-rock formations were thrust-faulted, fractured, and displaced in horizontal relationships so that huge masses of sedimentary rock tended to override each other. As this interaction continued, an anticline was formed in the advancing continental crust, an anticline similar to the Southern Rocky Mountains anticline described in the chapter entitled "Continental Divide." This anticline formed the Brooks Range from west to east, with the North Slope beyond it and a confused intermixture, a mélange, on its southern slope: enormous blocks of limestone shooting out at sharp angles, continental and ocean-bed rocks — a general mix of unrelated rocks. The North Slope was covered to great depth by the sediments from the Brooks Range anticline as it eroded, while the Canada Basin ocean crust tended to subduct beneath the northern edge to form the Barrow Arch. Then the subducting oceanic crust lost its initial vigor, and later the rifting ridge in the Canada Basin that had caused the original rotation ceased activity altogether.

The gross effect was the making of central Alaska out of Pacific Ocean bed and other rock together with the present sea-inundated Bering Strait off the Seward Peninsula. A large chunk of Siberia was also formed (the Chukchi Peninsula on the eastern side of the Bering Strait and the Chukchi Range, which is a continuation of a section of the rotated piece of North American craton). In addition to these events and at about the same time it is probable that the Arctic islands rotated into their present positions.

The net effect of these tectonic events was the formation of the Prudhoe Bay oil reservoirs. Many geologists, Gil Mull suggested, are puzzled as to how oil that was formed in post-Pangean Cretaceous shales got from impermeable shales into permeable Pangean rocks, the porous reservoir rocks of the prolific Prudhoe Bay oil and gas reserves. The explanation, Mull went on, is that the shales unconformably overlie Pangean sandstones because of the "skewed stack of playing cards" effect described above. As one goes northeast from Prudhoe Bay, the surface of the Pangean rocks, the reservoir rocks, dips from 2,450 meters to between 3,250 and 3,650 meters beneath the surface. It is a plunging surface 100 kilometers northward and eastward. Molecules of oil were squeezed up this 100-kilometer continuous-contact slope to be trapped at the highest point beneath Prudhoe Bay, the Barrow Arch. However, the weight of sediments on the overlying shales in which organic material was cooked and turned into oil and gas was not great enough to explain its comparatively fast maturation and migration into the Pangean rocks. It is believed that heat generated by rifting, thinning of the ocean crust, and related tectonics in the Canada Basin, which continued through the Cretaceous Period, contributed to the process.

Vitus Jonassen Bering, a Dane appointed by the Czar of all the Russias to lead an expedition to determine whether Asia and North America were connected by land, passed through the strait that now bears his name during August, 1728. Bering and his crew

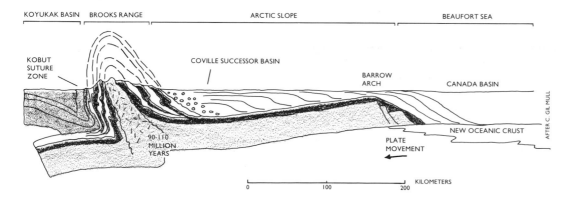

KOYUKAK BASIN   BROOKS RANGE   ARCTIC SLOPE   BEAUFORT SEA

KOBUT SUTURE ZONE   COVILLE SUCCESSOR BASIN   BARROW ARCH   CANADA BASIN

90-110 MILLION YEARS   NEW OCEANIC CRUST

PLATE MOVEMENT

AFTER C. GIL MULL

KILOMETERS
0   100   200

A spreading ridge in the Canada Basin beneath the Arctic Ocean is believed to have been responsible for rotating the Arctic Slope region of Alaska through an angle of 90° during a period between 135 and 110 million years ago.

didn't sight the fog-shrouded American shoreline. The first European expedition actually to see the North Slope of Alaska was that of the Yorkshire master navigator James Cook in the summer of 1778. But it was the intrepid Arctic explorer John Franklin, with his fondness for influential people, who in August, 1826, named Prudhoe Bay after Algernon Percy, first Baron of Prudhoe in Northumberland (locally pronounced, as Gil Mull stressed, Pruhdhoe, not Proodhoe). The Eskimo were the first to find oil seeps in the area and had probably used oil for lighting and heating purposes for generations before European oil exploration began; oil claims first were staked near Cape Simpson in 1921. A large number of people spent a great deal of time and multi-millions of dollars before one of the eleven exploratory wells drilled in the region over the years struck oil in commercial quantities. This was at Prudhoe Bay on September 26, 1967, at a depth of 2,400 meters and proved to be the largest *accessible* oil reserves ever to be found in North America. Discovering the oil was a major achievement. Getting the oil out of the ground in the extraordinary conditions of the North Slope was a miracle of determination. But providing the means of transporting oil by pipeline across permafrost tundra and several major mountain ranges to the port of Valdez in southern Alaska was a technological feat comparable only to the building of the pyramids of Giza or the landing of a man on the Moon.

It was now September, 1981, and it was snowing at Prudhoe Bay; the Arctic winter was about to extend its icy hand. At Deadhorse Airport Bob Hartzler of Sohio met Joy and me off the company's jet from Anchorage. He escorted us through the Base Operations Center, a veritable space station. To sustain the sophisticated life of 500 southern-latitude people in an Arctic clime requires the construction and the maintenance of a southern environment. Suddenly one understood why the dinosaurs that left footprints on rocks north of the Brooks Range had died out, why whales are insulated with blubber, and why the Eskimo are rotund.

The crew of the oil-drilling platform were delightful characters, for me epitomized by

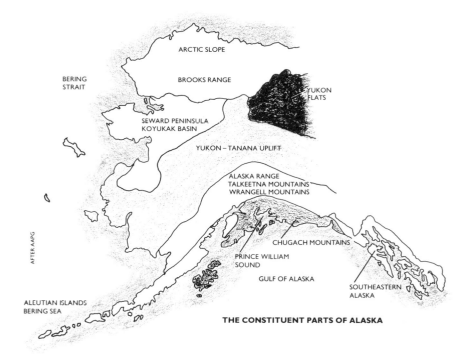

THE CONSTITUENT PARTS OF ALASKA

AFTER AAPG

one who wore a red bandana around his head and a gold ring in his nose. Right there in the noise and bustle of the rig's control room we were given a seminar on how oil wells are drilled. I was curious to know how holes are drilled like the ribs of an inverted umbrella from a single position on the surface — a trick that saves the need to drill from more than one position I learned that the bends in the spokes are very gradual, never more than 2.5° in 30 meters. The hole, 45 centimeters in diameter, is filled with drilling mud, which acts as a lubricant and which must be kept at a pressure greater than the subsurface pressure of any gas or oil or water in the rocks that the drill bit might encounter. If this procedure is miscalculated. there is a danger of a blow-back of explosive force. The direction of the hole is controlled by adjusting elements of the drill bit, and the geography of the hole is checked by replacing the bit with a survey tool. The survey tool has a compass, an inclinometer, and a camera built into it. The camera photographs the reading on the instruments. After the well has been drilled, logs are taken of the hole — ultrasonic logs, gamma-radiation logs, and so on — to provide a record of the drilled rock structures. The instruments

can measure the difference in electric potential for limestone versus sandstone or for gas-bearing rock versus water-bearing rock. From such information oil men construct a mathematical model of the hole.

When drilling is complete, the hole is lined with a steel sleeve, which is lowered down the

A number of wells at Prudhoe Bay are drilled from one point.

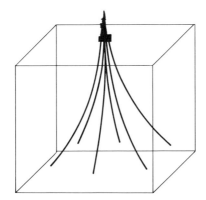

VALDEZ, ALASKA

The technical achievement of transferring crude oil from 2,400 meters beneath the Arctic Slope of Alaska at Prudhoe Bay and delivering it by pipeline to Valdez 1,280 kilometers to the south (panorama above) at the rate of 1.5 million barrels a day parallels the achievement of putting a man on the Moon. Once the source of the oil was discovered (the discovery well on the North Slope is pictured on the right), which took many years of high-cost effort, artificial means had to be developed to sustain a near-normal life for hundreds of people to work in Arctic conditions in all seasons. The Sohio Base Operations Center (pictured to the far right) provides that means and is the only full-scale building in permanent use on Earth that is comparable to the space station that will one day be necessary to sustain human life on the Moon.

SOHIO BASE, PRUDHOE BAY, ALASKA

hole 12.25 meters at a time and screwed together section after section to form a continuous pipe. The outer part of this sleeve is cemented to the bore hole, and the pipe is filled with diesel oil under pressure. It is controlled by a Christmas-tree valve arrangement at the top (see page 84, the discovery-well head). On the surface the line to the valve is filled with crude oil to prevent air from getting into the system. Explosive charges, shaped like bullets, are then lowered down the hole through a special valve. The charges are fired, perforating the metal and cement casing. The objective is to make holes in the sleeve where there are oil-bearing rock formations. At Prudhoe Bay these are 120 meters thick, but in other fields they can be considerably less, as little as 2 meters. There is usually water below the oil and gas above it; so the perforations must be made with care. The pressure on the diesel oil filling the hole during this operation is less than that of the oil in the reservoir rock so that, once the pipe has been perforated, crude oil can begin to flow from the reservoir rock toward the surface, where its flow is controlled by the Christmas-tree valve. The hole or, more correctly, the umbrella spread of holes is now an on-line oil well, which costs from $5 million to $20 million to construct. After completion of a well head the drilling rig is moved along rails to its next drilling location.

Crude oil straight from the well and champagne straight from the bottle have one thing in common: They both contain gas, which bubbles explosively when pressure is suddenly released. Oil from a well flows rather more copiously than champagne does from a bottle; to avoid the champagne effect, which would cause the oil to gush, pressure in the line must be slowly reduced to atmospheric pressure. The gas released from the oil during this operation is pumped back to the well along with sea water to replace the oil removed from the reservoir rocks and thus to maintain balanced pressure in the oil field. It is in fact possible to remove only about 40 percent of the oil from the rock reservoir; 60 percent remains in situ. There is a fortune waiting for anyone who can invent a way to extract a significant part of

this residue. Water is removed electrostatically from the oil, which is then pumped straight to the pipeline — one unbroken pipeline to Valdez, half of it above the surface and half of it below, ten pump stations, 1.5 million barrels a day over 1,280 kilometers southward.

At Tuktoyaktut, on the Mackenzie Delta, 500 kilometers due east along the Arctic shoreline from Prudhoe Bay, I had become acutely aware of the proximity of modern to ancient man. The contrast between technologies is overwhelming: the artificial islands of modern man in the Beaufort Sea and the nearby bone tools of ancient man in a river bank of the Old Crow in the northern Yukon. Here at Prudhoe Bay, figuratively just around the corner from the Bering Strait, I became vividly aware of the poignant juxtaposition of modern and ancient methods of survival, of the contrast between human ingenuity in the Space Age on the one hand and in the Ice Age on the other, one resulting in the exploitation of a major source of energy for today and the other resulting in the peopling of North America in prehistoric times.

The Bering Strait today is narrow, about 100 kilometers at its narrowest, a shallow seaway with often rough and treacherous waters that are frequently shrouded in mist in the brief spring-summer-autumn cycle and ice-filled or ice-covered throughout the long dark winter — not a hospitable place. And yet this narrow strait was once a focal point of life in the Northern Hemisphere. It was the approximate center point of the geographic link between the continents of Eurasia and North America. Scientists at first thought that the region was an occasional bridge between the continents. This has proved to be an understatement: It is now broadly accepted to have been not just a land bridge but a subcontinental region that was at times, throughout the last 150,000 years, perhaps 1,200 kilometers from north to south and about 2,200 kilometers from east to west. Archeologists call the subcontinent Beringia.

A comprehensive reconstruction of the general history of Beringia will take many decades to complete. But it seems that the subcontinent's geological history began when the cratonic arm of North America that formed the Alaskan Arctic slope was rotating through

90°. The arm formed the Chukchi Peninsula and Chukchi Range, which make up the Siberian Arctic slope — in many ways a duplication of the Alaskan slope. This relationship has been established by the matching rock formations beneath and on both sides of today's shallow Bering Strait and by the character of a suture between the Chukchi Range and the disparate rocks to the south of them. Eurasia and North America were joined for many millions of years; in fact here the geological connection is still as clear as that between the British Isles and the continent of Europe, Labrador and Newfoundland, or Connecticut and Long Island.

There was no seaway separating the North Pacific and the Arctic oceans until a temporary link was formed about 5 million years ago. This date has been established by the discovery of fossils of a particular species of North Pacific marine mollusc, which first appears in the North Atlantic fossil record less than 5 million years ago; the fossils were found in Tjörnes, Iceland. By 3 million years ago the molluscs were prolific on North Atlantic shorelines. These were warm-water creatures, and their presence not only confirms the establishment of an ocean link but suggests temperate climates in northern latitudes at the time of their dispersal from one ocean to the other. But the sea link was temporary because there followed four major periods of faunal exchange (mostly one way) between Eurasia and North America. During the period between 4.7 million years to 3.4 million years ago, bear, grison (weasel-like animals), deer, and panda migrated from Eurasia to North America, animals whose presence suggests temperate climates and wooded landscapes. About 2.5 million years ago lemmings and mole voles began to appear on the continent, and at 1.8 million years the first and most primitive of the mammoths. A change in climate had occurred that heralded the Pleistocene Epoch, the Ice Age. The most recent migrants, their arrival time measured in only tens of thousands of years past, were bison,

The term land bridge understates the true extent of the Ice Age isthmus between Siberia and Alaska: Beringia, the gray region of this diagram, was of subcontinental proportions.

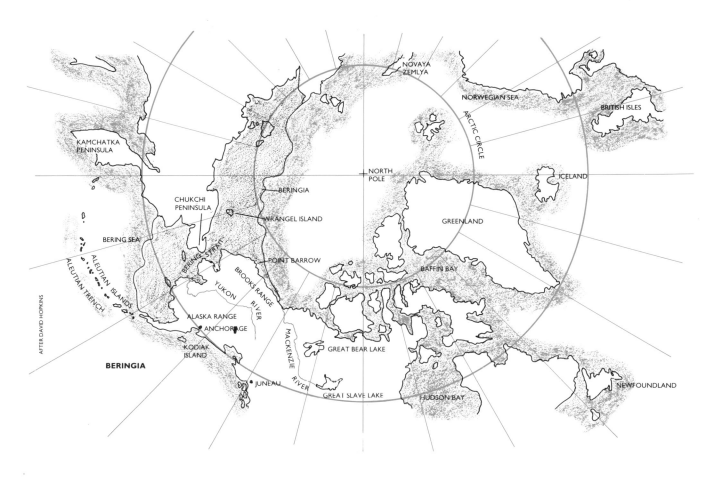

moose, caribou, musk ox, lion, mountain sheep, and *Homo sapiens.* Seventy-five percent of all land animals known to have inhabited North America migrated from Eurasia during this 5-million-year period of Earth's history.

At least eighteen cycles of glaciation have been recognized during the approximately 2 million years of the Pleistocene Epoch. Of these, four main periods of glaciation and four main periods of recession (including the present recession) are thought of as the classical periods of the epoch. During periods of heavy accumulation of ice on the continents, periods of maximum glacial advance, the level of the oceans was appreciably lower than it is at present, at least 85 meters lower at the maximum

of the last classical period of glaciation (the Wisconsin period) and 100 meters in the preceding one (the Illinoian). Over thousands of years the volume of water transferred from ocean basins to continental surfaces during peak episodes of glaciation was sufficient to expose and drain the shallow Bering Strait and in addition to drain huge areas of neighboring low-lying continental shelf to the north and to the south. During glacial intervals the ice cover decreased, and water moved back to the oceans to partially or wholly re-cover the continental margins with shallow seas.

At the height of periods of glaciation western Beringia (the Siberian element) was in a rain shadow, and in summertime central Beringia (the central Alaskan element) was in a stream of warmer air from equatorial regions.

These conditions produced a continental climate in Beringia, which attracted enormous populations of marine animals and birds, sea fish and shellfish, trout and salmon, and animals dependent on ice, walrus and varieties of seal, to Beringian southern shores. To the north, where landlocked Beringian ice bordered the gyrating Arctic ice, and to the east, where it met the continental ice sheet, the conditions must have been forbidding in the extreme. Beringian summers were warm and dry. Winters were cold, very dry, and almost snow free. At the height of glaciation oceans were colder, evaporation less, and general pre-

Seventy-five percent of all naturally occurring mammals on the North American continent migrated from Eurasia during the last 5 million years. Human migration is a recent phenomenon. The transfer from Beringia to the heart of the continent is believed to have been accomplished during interglacial periods through an ice-free corridor that periodically existed between the two enormous ice sheets that covered the continent (illustrated here).

cipitation therefore low. The landscape was Arctic steppe in character, mostly treeless and dominated by grasses, sedges, sage, and heathers. Aspen and larch grew in sheltered places; bison, mammoth, horse, reindeer, camel, and siaga antelope were among the larger animals present; and musk-ox, deer, and mountain sheep, the scarcer, smaller animals. It was altogether an easier terrain for people to hunt in and gather on than the present tundra is. Even so, it was an unforgiving land in which to survive.

There is some speculation among archeologists that there may have been a sparse but interrelated population of nomads in the Eurasian and Beringian Arctic steppe, and some argue that the idea of a bone culture as an alternative to a stone-tool industry should not be dismissed lightly. With few trees from which to build shelters the Beringians might have tried to use bone supports, such as mammoth tusks and rib cages, covered with animal skins for living quarters. And in turn this need would have promoted the skill of butchering animals with bone implements made from ready-to-hand materials: freshly splintered bone, which is sharp and naturally curved. But at the moment there is a Beringian enigma of major proportions. The oldest datings of artifacts in the heartland of Beringia, near the shores of the Bering Strait and the Bering Sea, are in the order of 12,000 years before the present. This seems to contradict the broadly agreed datings of 25,000 to 30,000 years before the present of Old Crow artifacts and apparently makes outrageous any suggestion for even older claims at Old Crow. One suggestion is that it is just too early in the Beringian archeological game to make judgments: Maybe techniques are not sufficiently advanced today to find and interpret the sites. After all, Beringia has only recently been recognized for what it is, a huge subcontinent that represents a great deal of future archeological digging. Beringian archeology today may well be where Egyptian archeology was before the discovery and interpretation of the Rosetta Stone.

In our own lifetimes we have not been conscious of a change in the level of the sea during the present glacial recession, and neither were

the nomads of Beringia in their lifetimes. They were not suddenly excluded from the area of the Bering Strait by inundation; the process of sea transgression and therefore enforced relocation of human beings was so gradual as to be unnoticeable. On the other hand I have been conscious of receding glaciers; the Rhône Glacier in Switzerland has receded noticeably since I first saw it thirty years ago. As glaciers receded in Beringian times no doubt the nomads also noticed, investigated, and probed the newly revealed territory.

Although the Brooks Range was glaciated, the mountains of eastern Beringia in the neighborhood of Old Crow (the Richardson Mountains) were not. But south of the Richardsons along the Rocky Mountain chain almost to Wyoming there extended an enormous ice sheet, the Cordilleran Ice Sheet. It was opposed on the other side of the Mackenzie River by a second vast ice sheet, which covered the Canadian Shield — the Laurentide Ice Sheet. At peak periods of glaciation the two met at their respective eastern and western extremities. But during the many glacial intervals — minor as well as major, hundreds and sometimes thousands of years apart — there was an ice-free corridor between them. It was through this 2,400-kilometer corridor (from the Mackenzie Delta to Calgary, Alberta) that nomadic tribes ventured and made their way to what they discovered to be a truly ice-free, lush, and well-forested region beyond — a biotic heaven after an icy purgatory.

After we had finished pondering Alaskan geology on the floor of the Captain Cook Hotel suite, I confessed to Gil Mull that I had a problem understanding why there should have been two major continental ice sheets and an ice-free corridor between them. "Then," Gil replied, "you must go south from Anchorage to Yakutat. When you have seen the glaciers in that region, you will understand exactly why there was a corridor on the eastern side of the Rockies. Besides, the place is a photographer's dream, and the regional geology is out of this world. Come to think of it, I insist that you go to Yakutat!" Of course I always do precisely what I am told to do — well, almost always, and then perhaps not quite precisely.

First I wanted to see the tidal bore on the Turnagain Arm off Cook Inlet. It is the way by road to Portage through the mountains to Whittier, from where one takes the ferry through Prince William Sound to Valdez, one-third of the way south to Yakutat. I stopped at The Bore Inn on the roadside beside the Turnagain estuary, sure that the landlord would know the tidal-bore times. Two ladies, the only occupants, sat with arms folded on the other side of a semicircular bar and looked at me speculatively as I entered. Somehow I didn't think that they could help, but rather than leaving without a word I asked whether either of them knew the time of the next bore. They looked at each other and then back at me, raised their arms, and in mock horror said something about its being none of my business. My surprise at their response must have equaled their surprise at my question; for I could now see that they were the topless denizens of a topless bar. I gave the bore a miss and got on to Whittier to catch the ferry to Valdez.

Prince William Sound, the 160-kilometer stretch of seaway between Whittier and Valdez, is a very extensive Pacific Ocean incursion into the Alaskan landmass. It was originally named Sandwich Sound (after the then First Lord of the British Admiralty) in May, 1778, by Cook, during his third and final voyage. Lord Sandwich later insisted that the name should be changed to honor his sovereign prince. Cook was under orders to seek the Northwest Passage from the Pacific to the Atlantic. The sound was the first of several promising passages en route to the Arctic Ocean that proved to be a cul-de-sac. The most misleading of these was Cook Inlet, now the site of Anchorage, and its offshoot, Turnagain Arm, in which Cook's ship, H.M.S. *Resolution*, had been unable to make way against the tide. Modern maps owe much to Cook, who probably made a bigger contribution to cartography than has any explorer before or since his time.

Today we have an advantage over Cook in his bewilderment at the intricacies of Alaskan inlets. We can stand back from a globe of the Earth and see Prince William Sound from a

The Earth's fractured upper crust, the lithosphere, is divided into plates that interact at their boundaries as they are moved in relationship with one another by dynamic forces beneath their surfaces. The plates more often than not consist of a reasonably balanced intermix of continental and ocean crust. The Pacific Ocean Plate is the great exception. The present Pacific floor is all that remains of the Panthalassa Ocean crust. It once surrounded supercontinental Pangea and consisted almost entirely of sediment-covered basalt with an occasional chunk of broken-off continental material embedded in it, the now-suspect terrains.

One form of interaction at plate boundaries is subduction, the plunging of heavier ocean crust beneath lighter continental crust. When this occurs, the subducting ocean plate interacts with the overlying continental crust to form a magmatic arc of volcanoes in the continental crust above it. When such an interaction takes place beneath the sea, between an ocean crust and a continental shelf, the volcanoes form an is-land arc. The Aleutian Islands are a supreme example of such an island arc and were formed as a consequence of the Pacific Ocean Plate's subducting beneath the North American Plate continental shelf.

WHITBY HARBOUR, YORKSHIRE, ENGLAND

THE SUTCLIFFE GALLERY

James Cook, considered by many to have been the greatest navigator of the age of heroic exploration, was captain of H.M.S. *Resolution*. This ship was a collier refitted at Deptford shipyard to (supposedly) withstand the rigors of Cook's historic voyages. Like Cook himself the vessel came from Yorkshire. It was built at Whitby, a small and still picturesque harbor on the east coast of England, where Cook had a house on the harbor until his untimely death in Hawaii in 1779. The picture of Whitby by a widely acclaimed local photographer, Frank Sutcliffe, circa 1880, captures the spirit of the sailing-ship era. The author's panorama contrasts the present-day Whitby Harbour with the locality of Prince William Sound in the Gulf of Alaska. Cook had almost to make a circumnavigation of the globe to reach the sound, and he explored it in 1778 during his endeavor to find the legendary Northwest Passage from the Pacific Ocean through the Arctic Ocean into the North Atlantic, thus obviating the need for the very voyage he had by then completed.

HMS RESOLUTION

tectonic perspective. The sound marks the North American end of an arc of islands, the Aleutian island arc, that starts off the coast of Alaska and sweeps westward across the Pacific to the Siberian peninsula of Kamchatka. On a map drained of the Pacific Ocean Prince William Sound can be seen to mark the eastern end of the Aleutian Trench, the trench into which the Pacific Ocean Plate is actively subducting beneath Alaska and Siberia; its movement is from south to north. The Aleutian Islands are an arc of volcanoes between the ocean and the continental shelf, which was once Beringia. The islands represent the zone of interaction between the upper crust and the ocean crust subducting beneath it. The trench is a plate boundary, and Prince William Sound is a focal point of one of the faults associated with that boundary.

Why an island arc? In fact why do many of the world's coastal volcanoes form arcs? It all has to do with the geometry of spheres. If one picks up a knife to slice a round object, the only way to form a straight line relative to the sphere's center would be to cut the sphere in half. Cut an off-center slice, and the edge of the slice will form an arc on the surface of the sphere. The thinner the slice, the smaller the radius of the arc relative to the sphere as a whole. The only way in which a subducting ocean plate could describe a straight line on the surface of the Earth would be for it to plunge vertically downward. At any other angle but the vertical the subducting plate can form only an arc. The radius of an island arc or that of a magmatic arc of land volcanoes on the edge of a continent is a good indicator of the angle of descent of a subducting plate: The smaller the radius, the shallower the angle of descent.

The surface regions of plate boundaries separate or collide or move by each other in jerky movements; and when they move, as at Krafla on the Reykjanes Ridge in Iceland, the movement is restricted to a limited part of the contacting plate boundaries. The movements are usually little more than a few meters at a time but occasionally 10 meters or more, and the periods of dormancy between movements can often be measured in hundreds of years. But when such a periodic movement takes place, that movement is sudden and violent. For ex-

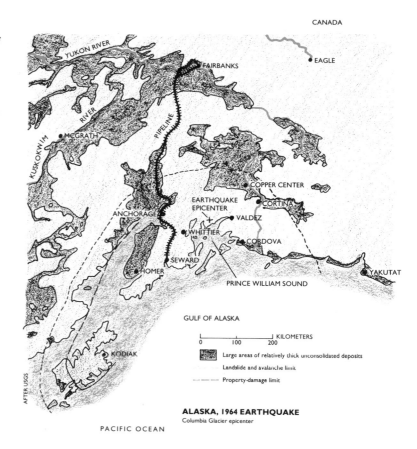

The epicenter of the Alaskan earthquake of 1964 was in the vicinity of Prince William Sound. The earthquake was the result of one pulse in the perhaps thousands of pulses which have been caused by the subduction of the Pacific Plate beneath the North American Plate.

ALASKA, 1964 EARTHQUAKE
Columbia Glacier epicenter

ample, late in the afternoon of Good Friday, March 27, 1964, the rocks associated with the boundary between the Pacific Ocean Plate and the North American Plate beneath Prince William Sound adjusted to the strain of subductive movement that had been building up in them at a depth of 20 to 40 kilometers beneath the sound. The relief of the strain was along the surfaces of a fault several hundred kilometers in length that runs parallel to the Aleutian Trench — near the edge of the subducting margin of the ocean floor. Relief of that rock strain caused one of the most severe earthquakes on record, the Alaskan earthquake. Although this earthquake was a human tragedy that cost 130 people their lives and the destruction of $500 million worth of property in a matter of minutes, in reality it was just one more pulse in the hundreds (maybe thousands) of such pulses that have taken place during the subduction of the Pacific Ocean Plate beneath Alaska.

Much of the destruction by an earthquake is caused by compressional shear waves. These are fast, short-wavelength vibrations (of high amplitude) that temporarily liquefy soils and cause mountainsides to collapse and structures to float away on their foundations as if on water. It is the low-amplitude waves that cause buildings to rock and crumble. In addition to vibrations the sudden vertical or downward movements of rock structures on the seabed during submarine earthquakes cause tsunami to form on the surface of the sea. The ripple patterns caused by a stone thrown into a pond are illustrative of tsunami waves in the sense that they radiate from the source of the energy that created them. Waves on a pond, and the tsunami also, reduce velocity and gain in

COLUMBIA GLACIER, PRINCE WILLIAM SOUND, ALASKA

NUNATAK GLACIER

BRAIDED STREAMS

The Columbia Glacier terminus (left), seen here in Prince William Sound, is a main tourist attraction and is visited by ship from Valdez or Whittier. The Columbia Glacier calves icebergs constantly because it flows into a deep fjord. Although it is receding at about 500 meters a year, geologists consider it to be still in the advanced position achieved during the last "little" Ice Age of 3,000 years ago. However, USGS computer-model predictions suggest that by 1985 there will be a further 10-kilometer reduction of the present 67-kilometer length of the glacier.

A second sea glacier, Nunatak Glacier, is pictured at the bottom left. This shows that the terminus of this glacier has also retreated very substantially in recent times. The old surface level of the glacier can be seen marked on the cliff on the left of the picture.

As glaciers recede, the bottom of the valleys they have cut are filled with fluvial deposits that are distributed evenly over the valley floor by braided streams (picture at the bottom right). If glaciers continue their present retreat, such newly formed valley floors will become covered with vegetation.

height as they reach shallow water. The difference is that the tsunami are high-speed waves produced by enormous pulses of energy. In midocean they can travel at 700 kilometers per hour at a height of 0.25 meters with a wavelength (distance between crests) of about 95 kilometers; such waves gain 25 meters in height at the reduced speeds inevitable as they arrive in shallow water. Tsunami inflict immense *further* damage to shorelines that have already been wrecked by low- and high-amplitude vibration near the epicenter of an earthquake and damage hundreds and perhaps many thousands of kilometers away from the source. The shores of Prince William Sound

were affected in this way during the Alaskan earthquake, including those in Valdez harbor at the Alaskan Pipeline terminus.

The character of the mountains immediately east of Valdez, the Wrangell Mountains, has been attracting a great deal of attention from geologists during the last few years. Although the mountains are intermixed with volcanics, as one would expect in a region so near to the subducting Pacific Ocean Plate, Mount Wrangell and its companions are actually alien to the North American continent. They are immigrants from the single ocean that surrounded Pangea — the Panthalassa Ocean. They were conveyed to their present locations by the movement of the proto-Pacific Plate. Whence they came is a matter of speculation; they and much of the present western

margin of the North American continent are in fact "suspect terrains." The professional shorthand for rocks of foreign origin is "allochthonous," but since this word is almost unpronounceable, I'll stick to suspect terrain.

David L. Jones of the USGS, Menlo Park, and his colleagues are "hunters of the Snark" in the sense that Pacific suspect terrains and their origin are their scientific domain. I visited Jones in his laboratory in California to glean more about this new and fascinating concept, which I had first heard about from Fred Dunning of the Geological Institute in London. Dave Jones is a man of great energy and enthusiasm for his subject. He bounded about from wall map to wall map.

MOUNT WRANGELL, ALASKA

During recent years it has been shown that much of the coastal region of the West Coast of North America from Alaska to California west of the Rocky Mountains consists of huge pieces of continental crust and oceanic crust that did not originate in the locality in which they are now found. These new elements were rafted into their present position by movements of the Pacific Ocean crust. It is believed that the only parts of the present continental margin that are "homemade" are the volcanoes and associated volcanics. This truly amazing discovery of what are termed suspect terrains means that on the Pacific West Coast we are actually seeing the making of a continent in progress.

The picture at the left is of Mount Wrangell, east of Prince William Sound. This mountain, together with its associated range, is one of the suspect terrains summarized in a very general way on the map at the right.

This diagram illustrates how the rafted rocks, such as those of Wrangellia, are deformed on collision with the continental margin.

When Pangea had fully formed and before the Tethys Sea began to separate the huge continent into the future northern and southern continental elements, there could have been only one ocean, the ocean called the "Panthalassa," and this of course surrounded the supercontinent. But when the Atlantic Ocean began to form, the Panthalassa Ocean began to shrink correspondingly and thus became the proto-Pacific Ocean. Although the concept of Pangea as one vast continent is unaltered, recent discoveries suggest that in fact there were microcontinents in the Panthalassa Ocean that were unattached to Pangea. It is also thought that large pieces of the extensive coastal regions of Pangea moved around the supercontinental coastline. Some of these pieces became islands far removed from their places of origin — sometimes by many thousands of kilometers. Some attached themselves to parts of the Pangean coastline to which they did not belong. This migration continued throughout the breakup of Pangea and the development of the Pacific Ocean. To the east of a general line of faulting and west of the Rocky Mountains huge areas formed during the Laramide orogeny are suspect: They do not belong to the continental region of which they are now part. Their geology and often their fossil record simply do not match those of the North American craton. The process of rotation, microplate tectonics, is continuing today in the Pacific Ocean. New Zealand, for example, is a displaced segment of continental material: At the moment no one is able to identify its origin. Alaska is mostly an assemblage of a number of large pieces of microcontinent. In fact the whole of the West Coast of North America is largely an accretion of suspect terrains. Some of these are being driven along the coast by the present movement of the Pacific Plate. Captain Cook was seduced into deep inlets that looked to him like seaways to the Arctic Ocean and the Northwest Passage. In fact Cook's inlets are there partly because large microplates have not yet been rammed home by Pacific Ocean Plate movement, to fit snugly and mountainously against an accumulating continental mass.

All the shaded territory is terrain that has moved into its present position either from an adjacent region (like the Arctic slope) or from afar. Other continental scraps that match the Mount Wrangell rock sequences are distributed southward, and these are termed Wrangellia.

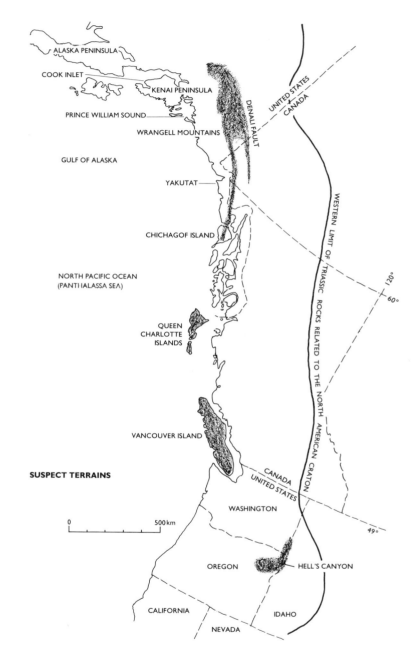

HUBBARD GLACIER NEAR YAKUTAT, ALASKA

The Hubbard Glacier is one of many immense glaciers in southern Alaska near Yakutat, where moist Pacific air constantly precipitates on the surrounding high peaks formed by the collision of suspect terrain with the continental margin. The ranges of mountains that can be seen in the panorama include Mount Logan (6,050 meters), the highest in Canada (at the left of the picture).

Over the horizon, days are sunny, mostly dry, and frequently warm and snow free. The valleys over there are in a rain shadow and thus benefit from the heavy precipitation on the mountains to their west. So it was during the latter stages of the Ice Age that mankind migrated from Beringia to continental North America.

Then these mountains were almost buried in ice just as Greenland is buried today. Beyond the mountains there were valleys in the vicinity of the Mackenzie River and its tributaries where precipitation was low. Since glaciers form more frequently on warmer western slopes, such as those in the picture, ice was low and even nonexistent during intermediate stages of glaciation to the east.

Through this corridor the first people to populate the central regions of North America gained entry to the lush, comparatively warm, and often well-forested plains beyond the glacial-outwash regions.

This is a region of perpetual snow, on sunlit slopes in summertime perhaps very wet snow. This panorama was taken above the Hubbard Glacier. The large mountain to the left is Mount Vancouver (4,800 meters), which marks the boundary between the U.S. and Canada (Alaska and British Columbia).

At the bottom center the glacier that curves upward and off to the right toward a snowfield is Variegated Glacier, a "surge glacier," which can suddenly move with a wave velocity of from 150 to 400 meters per hour. The cause of such surges is thought to be the accumulation of a cushion of water beneath the glacier. When the pressure of the water is sufficient, it lifts the glacier free of basal obstructions, allowing the glacier to slide rapidly downslope.

VARIEGATED GLACIER, ALASKA

OCEAN CAPE NEAR YAKUTAT, ALASKA

Ocean Cape at Yakutat in southern Alaska. Immediately east of here the moisture-laden air of the prevailing winds from the Pacific Ocean meets the snow-covered peaks of high mountains, which rise dramatically from sea level to a maximum near 6,000 meters. As a result of this interrelationship Yakutat is a region of extremely high rainfall, evidenced by the rain forests that flourish along the bar of fluvial deposits separating the Pacific Ocean from the estuarine environment at the left of the picture.

The fast-flowing gravel-bottomed streams beyond the estuary are very attractive to varieties of Pacific salmon, which return here from great distances after spending most of their lives in the ocean. On their return to the estuary first they undergo changes in color and even appearance as they adjust to the freshwater environment. Then they make their way back to the very streams in which they were first hatched; here they spawn and die. The young hatch out as the ice melts the following spring and remain in local streams, sometimes for several years, before returning to the Pacific Ocean to recommence the cycle.

Every animal in the Yakutat community (including man) directly or indirectly depends for its living on the existence of the salmon — bear, wolverine, sea otter, and bald eagle, to name a few of a great number.

This startling concept brought me a new sense of excitement at the prospect of seeing the magnificent mountains that rise majestically and to considerable heights out of the Pacific for thousands of kilometers southward along the Alaskan coast, the Canadian Pacific coast and the West Coast of the United States. The volcanoes in these regions were formed by interaction between the ocean crust and the continental crust. But most of the mountains are imported assortments of rocks from South America, from Mexico, and from unknown sources in the Pacific Ocean itself. These rock formations — sedimentary rock, metamorphic rock, and igneous rock from the ocean floor — have been tilted, folded, and generally abused in the process of their transfer to, and eventual amalgamation with, the edge of the North American continent.

I had been well briefed about the Yakutat region by Gil Mull, and for this reason I was able to take maximum advantage of breaks in otherwise constantly inclement weather in this extraordinary area. Here in 1774 the Spanish began their brief brush with northern climes, and here also Vitus Bering and later Captain Cook had their first views of Alaska.

In the vicinity of Yakutat unobstructed moisture-laden winds from the vast Pacific Ocean are suddenly confronted by some of the highest mountains in North America: the Wrangell, the Chugach, and the St. Elias mountain ranges (Mount Wrangell, 4,350 meters; Mount St. Elias, 5,530 meters; Mount Logan, 6,030 meters). The moist air is forced high by the mountains, and this results in almost perpetual heavy rain on the coast and snow in the mountains. The result is glaciation on a scale unprecedented other than in Antarctica and Greenland. Massive erosion in the region is the consequence of relatively high coastal temperatures and the protrusion of glacier snouts into the sea.

The glaciers here are innumerable and vast: The Malaspina Glacier, for example, is 80 kilometers wide and reaches back 50 kilometers into the mountains; it is a remnant ice sheet of the Wisconsin glacial advance. I was looking for a glacier that would illustrate by its dimensions at sea level and its contrast in height with surrounding mountains why there had been an Alaskan corridor through which ancient man could have passed during the interglacial stages of the Ice Age. Gil Mull's recommendation had been Hubbard Glacier, but the problem was how to photograph a panorama of it. After a lot of searching from a fixed-wing plane above the glacier I decided that the only solution was to arrange a helicopter landing on the top of Gilbert Point (1,176 meters), overlooking the Hubbard. After ten days of waiting for a break in the weather one evening Joy and I were dropped off with the promise by the pilot that he would return two hours later — if the clouds had not closed in and it wasn't too dark! Fortunately the tentative emergency arrangements for an enforced stay that we had made were not necessary, and the picture on pages 96 and 97 was the result of a hectic scramble to unpack, set up, take the panorama, and repack before the weather closed down or the helicopter returned. It was quite an experience. Perhaps the most spectacular view of a lifetime, but ". . . full of care, we had no time to stand and stare," as W. H. Williams proclaimed in his poem "Leisure."

ACTIVE MARGIN

AFTER a day's trolling in their fishing boat, *Edrie*, Howard G. Ulrich and his seven-year-old son had negotiated a strong tidal current through the narrow entrance of Lituya Bay, to anchor for the night in a small cove. Two other fishing vessels were anchored in another cove on the other side of the inlet. It was a clear, balmy, July evening on the south coast of Alaska, about 160 kilometers south of Yakutat. The setting is idyllic. Lituya is a deep natural harbor about 10 kilometers long and 2 kilometers wide, with Cenotaph Island at its center. Less than 1,000 years ago the mouth of the bay was the terminus of Lituya Glacier, which has now receded out of sight around the far left-hand headland of the T-shaped inlet. The snow-covered peaks of the Fairweather Range loom high and abrupt on the horizon beyond the headlands. Unknowingly, the Ulrichs in their boat and their overnight companions, the crews of the *Badger* and the *Sunmore*, had been snared in a tectonic trap.

According to Don J. Miller of the U.S. Geological Survey (USGS), it was still light when a few hours later the *Edrie* reacted violently to undulations caused by major earthquake tremors in the bay. The tremors were followed by the sound of avalanches in the mountains. Suddenly, after a "deafening crash at the head of the bay," there appeared between and above the two headlands an explosion of spray intermixed with large blocks of ice. From beneath the spray there emerged a wave of unbelievable proportions. The side of one of the Fairweather Range mountains above the Lituya Glacier had collapsed into the fjord below, and 90 million tons of rock had plunged into the water. The ensuing wave swept 525 meters high over the edge of the left-hand headland directly opposite and 183 meters high over the edge of the right-hand headland. The sailors in the three boats were faced with a gigantic wall of water racing toward them. It stretched from one side of the inlet to the other and well beyond its shores. The *Edrie*'s anchor chain snapped as she rose to the wave, and the boat and the Ulrichs were flung violently over the nearby shore to be returned into the bay by the backwash of the wave. The *Badger* was carried over the spit that protects the inlet from the Pacific Ocean and was wrecked, but her crew survived. The *Sunmore*, which had managed to get under way and had reached the entrance to the inlet, was swamped. Her crew were lost.

Had Lituya Bay been the site of a metropolis like San Francisco (and the localities have much in common), the human tragedy would have been multiplied perhaps a thousandfold. In fact the main victims were trees: millions upon millions of shoreline trees, which were uprooted, stripped of their branches, and denuded of bark in a flash by the turbulent power of the wave and the violent collisions of the trees with each other. Many of the logs that were swept out to sea accumulated at Ocean Cape, Yakutat (see pages 98 and 99). Although nearly a quarter of a century had passed since that summer's evening in 1958, the extent of devastation and its clear demarcation was easily evident to me from the air. I had no difficulty in imagining the huge ice blocks that had been left high on the shores above the bay; for a thousand meters of ice had been lopped off the full width of the Lituya Glacier terminus by the rock fall. The bay itself was filled with mush ice and thousands of trunks that floated on its surface. I could easily make out the scars on the mountain face left by the slide. And according to Miller, this wave was only one of a number of great waves that have previously devastated the region.

Beyond the Lituya Bay headlands there is a deep trench at right angles to the inlet. The trench runs straight as a die north for several hundred kilometers toward the Hubbard Glacier and the St. Elias Range, where it is presumably buried under the enormous glaciers of that region. The trench, the Fairweather fault, separates the mountains from the Lituya foreland and is partly filled in this region by a succession of glaciers fed by subsidiary glaciers from mountain valleys at right angles to it. Both the Fairweather Range to the east of the fault and the less formidable Lituya foreland to the west of it are suspect terrains. The Fairweathers plus the St. Elias and the Wrangell mountains to the north of them have all traveled great distances (the Wrangells are thought by some to have traveled as much as 9,000 kilometers from an equatorial latitude) but are now established parts of the North American continent. In fact some geologists believe that these terrains, and other provinces like them, which are plastered onto the edges of continental masses surrounding the Pacific

Ocean, may be remnants of early Panthalassa continents. However this may be — and this last suggestion is a highly controversial thesis — the foreland of southern Alaska in the vicinity of Lituya Bay is still edging its way along the coast, propelled by the Pacific Plate. On the day of the giant wave the Lituya foreland moved 7.3 meters northward. Eventually it too will dock with the newly accreted continental mass that constitutes most of Alaska and 70 percent of the Cordilleran region of North America.

The Fairweather fault continues in a direct line from the head of Lituya Bay southeastward beneath the Pacific Ocean. It runs beneath the sea until it joins a spreading ridge in the ocean floor that stretches from a point off Vancouver Island to a point off Cape Mendocino, California. The submarine Mendocino fracture zone, which intersects this point (in what is called a "triple junction"), is the connecting structure between the spreading ridge and the San Andreas Fault, which emerges above the ocean at Point Arena, 2,000 kilometers to the south of Lituya Bay. The San Andreas Fault is a counterpart of the Fairweather fault. Both are called "transform faults" because they transform into other structures — in this case spreading ridges. Both faults are surface manifestations of the boundary between the Pacific Plate and the North American Plate. They are faults that simply translate the northwestward movement of the Pacific Plate into surface reality: Anything to the west of the faults moves northwestward. The continent generally is advancing from east to west with little lateral movement. In relative terms the continent is at rest, and the ocean crust moves transversely northwestward along its western edge. That 2,400-kilometer-long piece of suspect terrain west of the San Andreas Fault, which stretches from Point Arena to the southern tip of Baja California, is simply being rafted along the coast.

The average northwestward movement rate of the Pacific Plate — and therefore of suspect terrain rafted upon it — is 5.3 centimeters a year. Deep beneath the ocean plate's boundary with the continent, about 15 kilometers below the surface, this movement is probably smooth because of the relative plasticity of the interfaces of both plates. But the nearer to the surface, the cooler and less plastic the interfaces become and the more rigid and inter-

LITUYA BAY, ALASKA

Lituya Bay, on the southern Alaska coast 160 kilometers south of Yakutat, has several times been the site of giant waves. One wave is known to have been considerably higher than the world's tallest building, the 440-meter Sears Building in Chicago.

On the evening of July 9, 1958, an earthquake caused the side of a mountain beyond the left-hand headland (the right-hand mountainside in the small picture at far left) to fall into the fjord beneath. A giant wave was generated by the sudden impact of 90 million tons of rock on the fjord. The wave swept over the left-hand headland to a height of 525 meters and 183 meters over the right-hand headland. Millions of trees were swept from the mountainsides, and a trim line marks the maximum heights reached by the wave (on the left of the same picture). The wave also destroyed forest well back from the shorelines, and here again the trim line marks the extent of the wave's impact (picture at near left).

The foreland in the panorama is suspect terrain: a large area formed in some distant place and being moved by the Pacific Plate along the Alaskan coast. The 1958 earthquake was caused by a 7.3-meter movement of the foreland from the right to the left of the panorama along the Fairweather fault on the horizon beneath the cloud cover. The Lituya Bay foreland's move is analogous to the movement of the Californian coastal region along the San Andreas Fault, 2,000 kilometers to the south.

THE SAN ANDREAS FAULT

POINT ARENA, CALIFORNIA

The coastal area west of the San Andreas Fault, the San Andreas block, is a suspect terrain that during the last 30 million years has been rafted about 500 kilometers northwestward from the position in which it was assembled from a number of unrelated pieces. The transverse movement of the San Andreas block relative to the mainland is caused by the movement of the Pacific Plate "conveyor belt." When deep sub-surface movement of the plate is translated to surface movement, earthquakes inexorably occur in the coastal regions of California adjacent to the fault. The longer a major surface movement is delayed, the greater the buildup of stress on opposite sides of the fault and the greater the resultant earthquake.

Human beings seem to have a lemming-like drive to live, work, and actually make use of regions on the line of the fault: Arterial roads follow the San Andreas for hundreds of kilometers; bridges, intersections, and homes have been erected in its neighborhood; reservoirs in the San Francisco area have been built in depressions formed by past movements of the fault. However, the buildings in which earthquake research is conducted by the USGS at Menlo Park, east of the fault, have been reinforced by steel girders of such quantity and strength that it leaves one in no doubt of the occupants' belief in their own predictions.

Near Point Arena (left): The San Andreas Fault rises to the land surface from the seabed in this region. The fault is in fact the boundary between the Pacific Plate and the North American Plate. In the picture the severely tilted terrain west of the boundary is moving northwestward at an average of 5.3 centimeters per year.

Southwest of San Francisco (below, left): Parts of Daly City perch on the very edge of an insecure cliff near Mussel Rock (right-hand edge of the picture). The San Andreas Fault follows a northwest-to-southeast line beneath the sea from Bolinas Bay to this point, where it emerges after crossing the Farallones Gulf west of the Golden Gate.

About two-thirds of the way from San Francisco to Los Angeles the fault, which maintains an almost undeviating straight line from Point Arena to this point, is 65 to 85 kilometers inland. Fault movement here is most apparent on the Carrizo Plain (below, center), where stream beds have been displaced or abruptly cut off by lateral movement of the block to the right of the picture.

The California Aqueduct (below, right) between Frazier Park and Palmdale, near Los Angeles, runs parallel with the San Andreas Fault. Here the fault follows a line to the southeast toward the horizon between the two ridges at the top right of the picture and so continues to join the East Pacific Rise at the mouth of the Gulf of California.

DALY CITY

CARRIZO PLAIN

CALIFORNIA AQUEDUCT

Point Reyes headland is a spectacular example of a displaced, or suspect, terrain; geologists use the word allochthonous to describe this and similar rocks of doubtful origin. They were formed or produced elsewhere and have been rafted into their present position by the present northwestward movement of the Pacific Plate.

POINT REYES, CALIFORNIA

locked they can be. The smooth subterranean movement gradually translates into sudden unlockings of fault interfaces near the surface when sufficient strain upon them has accumulated. Sudden relief from great stress results in sudden large movement between previously interlocked faces. In Lituya Bay the consequence was the 7.3-meter northerly movement of the Pacific Plate in 1958 (the foreground of the panorama on page 103). In the region west of the San Andreas Fault, 300 kilometers southeast of San Francisco Bay, the relief caused in 1857 a 10-meter northwesterly movement — much greater than the 6.5-meter movement that resulted in the San Francisco earthquake of 1906. In some regions, perhaps because of the character of the rocks near the surface, interfaces may not be completely interlocked, and in such instances

stress is relieved by more frequent and less severe movements and therefore less severe earthquakes. It is in these regions of the San Andreas Fault that artificial efforts to lubricate the interfaces of the fault are believed to be possible. But it seems that little can be done in areas such as Palmdale, 75 kilometers northeast of Los Angeles, where the strain along 250 kilometers of the firmly interlocked surfaces of the San Andreas Fault has become acute. For the last twenty years accumulation of stress has been manifested in this area by what geologists call the Palmdale bulge, a several-centimeter-high uplift along the line of the fault. It is here that the next major adjustment of the San Andreas Fault between the Palmdale bulge and the Gulf of California through San Bernardino is anticipated.

Of all the suspect terrains along the Pacific coast none could be more interesting than that part of North America west of the San Andreas Fault which includes the Big Sur coast, Los Angeles, San Diego, and the Baja Peninsula. I was with Dave Jones, of the USGS,

Menlo Park, learning something of his current work on suspect terrains, when one of his associates came hotfoot into the room from a field trip to Baja California, where he had just identified, so he announced blithely, "Peruvian-like terrain." The comment caused neither consternation nor disbelief. The non-reaction to such a startling statement tended to emphasize for me the sudden state of flux in the field of microplate tectonics as a result of recent discoveries along the full length of the North American Cordillera. The possible discovery of a bit of Peru in Baja California added more to the puzzle than it contributed to the solution.

A few years ago, I gathered, most geologists accepted that California west of the San Andreas Fault has moved northward up the Pacific coast by about 500 kilometers during the last 30 million years. But all agreed that there were inexplicable anomalies in the region, differences in rock types and ages, fos-

sils that didn't match other fossils in neighboring formations — discrepancies that could not all be explained by 500 kilometers of movement. But while the concept of continental accretion by rafted terrains, for which Jones and his colleagues are largely responsible, was being established, some of the anomalies in California were reappraised in the light of the new theory. It was then realized that the entire block west of the San Andreas Fault, the San Andreas block, which makes up the western edge of California and Baja California, could well consist of an accretion of suspect terrains. For example, at Pigeon Point about 65 kilometers south of San Francisco, there are shoreline rocks that seem to correlate perfectly with a region of the Sierra Madre del Sur, about 2,500 kilometers southeast on the coast of southern Mexico. The paleomagnetic characteristics of the rocks in both places strongly suggest that Pigeon Point did indeed raft through this extraordinary distance and that the transfer had been almost completed before 50 million years ago. But other parts of the San Andreas block appear to have had differ-

ent origins. Investigation has suggested that the whole San Andreas block amalgamated in one place lower down the Pacific coast. Then, about 30 million years ago the block began to move the 500 kilometers northwestward along the line of the San Andreas fault into its present position. This last movement satisfies the conventionally accepted movement of the coastal region, and prior amalgamation provides an explanation for the anomalies.

There are many more instances of a similar kind of transfer along the whole of the Pacific coast and inland from the coast sometimes to considerable distances. How does one respond to these claims, I wondered. "Ask three people about the origin of some of these terrains and you will get three answers," Jones responded. "In ten years we might have sorted it out — but that's being optimistic. All that can really be said for certain today is that we

have an extremely complex problem; a great and exciting geological mystery to solve."

The geological ages of the rafted terrains that are now known to constitute most of the North American Cordillera west of the Rocky Mountains, the time of their rafting on the ocean plate, and the time of their docking with the North American continent cannot have been coincident. The movement and accretionary processes were spread over hundreds of millions of years. Some docking of suspect terrains took place before the full assembly of Pangea, some during its existence, and in a sort of sweeping-up operation considerably more docking after the breakup of Pangea. Some pieces of suspect terrain are of continental origin; they contain cratonic rocks. Other terrains are recognized to be ocean-floor elements: seamounts, island arcs, and ocean crust. With this scenario in mind one wonders how the original North American continental margin now adjacent to the Pacific Plate, the Precambrian basement and the sta-

The rafting of large chunks of terrain on presently moving ocean floors over sometimes considerable distance is not by any means restricted to the coast of California or to the Pacific coast generally; it is a worldwide phenomenon. For example, the strikingly similar panoramas shown here are geographically 10,000 kilometers apart.

The upper picture is of a sea inlet, Tomales Bay, which separates "suspect" Point Reyes to the right from the Bolinas Ridge on the Californian mainland to the left. We are looking southeast down the line of the San Andreas Fault, which divides the two. The region to the right of Tomales Bay on Point Reyes island is Inverness Ridge, and the small town about halfway along the inlet is Inverness.

The lower picture is of the Great Glen, which separates the North of Scotland (to the left) from the Grampian Mountains (to the right). We are looking northeast along the Great Glen fault, with Loch Ness below and the original town of Inverness on the east coast of Scotland on the horizon. The whole of Scotland north of the Great Glen is suspect terrain, which has moved from afar into its present position.

ble platform secure upon it, reacted to the tectonic traumas of the times. One reaction was the formation of the Sierra Nevada, the Coast Ranges east of the San Andreas Fault, and the great Central Valley of California between the two.

About 135 million years ago the North American continent was being propelled in direct opposition to the spreading Panthalassa Ocean floor, the Pacific Ocean to be. This movement was caused by some unknown change in the Earth's dynamics that caused Pangea to begin to break up. The first event was the rifting and formation of the Tethys Seaway, which divided Pangea into Laurasia to the north and Gondwanaland to the south. This was followed by the formation of the Mid-Atlantic Ridge and the proto-Atlantic Ocean. For a time at least *all* continents moving away from the Pangean community on the sphere of the Earth could move *only* in opposition to a surrounding plate of oceanic crust.

At this time the margin between the North American craton and the Pacific Ocean crust was far removed to the east of the present margin. The ocean crust then in contact with the continent is called the Farallon Plate. The San Andreas block had not been accreted, much less moved into its present position. The Farallon Plate 135 million years ago was moving eastward in opposition to the continent, which was moving westward. The result was violent interaction between the converging crusts. Not only did the ocean plate subduct at a shallow angle beneath the continent, thought to have been about 7°, but the nose of the continent ploughed up the surface of the ocean plate to form a mélange, a nonhomogeneous mix of continental rock and ocean-floor material. The friction caused by interaction between the ocean crust and the continental crust about 100 kilometers below the surface was sufficient to melt the rocks to form what are called "plutons." A multiplicity of plutons joined together to form a "batholith." The plutons were up to 80 kilometers long by 25 kilometers wide. The Sierra Nevada batholith is composed of hundreds of plutons up to this size, which accumulated over a period of 60 million years to form an Andes-type mountain range. The Sierra Nevada of today is an eroded remnant of the resulting original mountain range.

Whether plutonic magma on continental margins forms in curtains or in blobs is not known, but it is known that plutons gravitate toward the surface. Being very hot, they are lighter volume for volume than the rock that surrounds them. Plutons form steep-sided chambers of semiliquid magma, which eventually crystallizes as granite from the top of each chamber to the bottom. During this process, as in the crystallization of all substances, water is released as a necessary by-product. In plutonic magmas minerals that may have been transferred from the subducting ocean crust into the liquid magma or that may already be present in the original continental crust concentrate as metal-rich compounds in the water to form hydrothermal solutions. It is these super-saturated solutions which are injected into fissures and other weaknesses in the granite matrix as it solidifies. Different combinations of minerals in these solutions at different levels and at different times in the plutonic cycle precipitate to form copper, silver, lead, tungsten, and gold deposits in a variety of forms and degrees of richness.

During the formation of the Sierra Nevada the upward progress of plutons partially melted and crowded aside pre-existing rocks. Volcanoes began to erupt on the surface, forming a volcanic arc. When magma reached the surface, those plutons with upper regions still in a semiliquid state exploded in fantastically violent eruptions, the like of which have not been experienced on Earth in modern times. The surface of the surrounding land was repeatedly buried by glowing avalanche flows of volcanic ash hundreds of meters thick. As pressure was relieved by surface eruption, granite could form in the upper parts of the plutons, about 4 kilometers below the surface. The process of plutonic streaming from the subduction zone continued during cooling and crystallizing. This resulted in crowded amalgamation of plutons and the formation of a batholithic mountain range.

As the Sierra Nevada batholith was forming, the edge of the continental crust beneath the sea was nosing out along the ocean floor, churning up deep sedimentary rocks and ocean-crust material. The resulting mélange piled up into a confusion of rocks that formed much of the Coast Ranges, leaving a trough

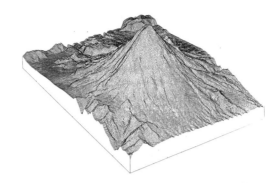

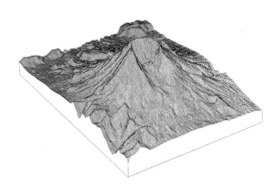

The eruption of Mount St. Helens on May 18, 1980, was caused by the present subduction of the remainder of the Farallon Plate (now called the Juan de Fuca Plate) beneath Oregon and Washington. This micro-plate is all that is left of a huge ocean plate, the Farallon Plate, which lay east of the East Pacific Rise spreading ridge, the Pacific Ocean equivalent of the Mid-Atlantic Ridge. Subduction of the micro-plate caused the formation of the Cascade Range magmatic arc. This arc of volcanoes includes Mount Rainier, Mount Hood, and many other large volcanoes in addition to Mount St. Helens. In terms of millions of years the intensity of subduction will modify as what is left of the plate is consumed beneath the continent.

The perspective views of Mount St. Helens reproduced here were made by the USGS Western Mapping Center. They are three-dimensional computer models (digital elevation models) of the mountain as if it were viewed from the northeast before and after the devastating 1980 eruption.

This picture is of a mineral mine at Idria, west of Coalinga, California. The Romans called the mineral *hydrargyrum*, "liquid silver." Our name for it is mercury, after the Roman god of merchandise and science. Mercury was mined in Spain at Almadén and in California at first at New Almaden. The metal is about 100 times rarer than copper is and occurs in rocks as an ore that contains on the average about half a gram per ton. Like other metals it is primarily a product of magmatism. Occasionally mercury is found in its native state, and this mine at Idria in the Coast Ranges of California is one of those places. The exposed galleries of the mine on the mountainside have a silver-gray look about them, which is due to rock altered to asbestos, a mineral often associated with mercury in this region. Mercury is separated by roasting the rock and by condensing the vapor given off.

THE NORTH PALISADE, SIERRA NEVADA, CALIFORNIA

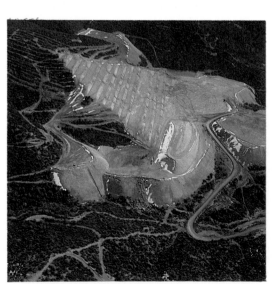

MERCURY MINE, IDRIA, CALIFORNIA

The North Palisade (4,350 meters). There could be no better example of the influence of geology on the destiny of life than these granite mountains. They were formed as a consequence of the North American continental edge's moving westward over the Farallon Plate, which subducted beneath the continent. Interaction between the two plates resulted in the melting of continental rocks above the subducting ocean plate. This produced huge plutons, semi-molten granite masses, that joined together to form one gigantic mass, a batholith. Today this batholith forms the backbone of California, and in its eroded form it is called the Sierra Nevada.

The great height of the Sierra Nevada interacts with the continuous flow of moist air from the warm Pacific Ocean and causes extremely high precipitation of snow on the mountain tops and rapid erosion of their sunny western flanks. Melting snow from the mountains carried alluvium down by streams and rivers to fill the shallow sea basin between the Sierra Nevada and the Coast Ranges to form the present Central Valley floor. The tributary streams also transferred placer gold to the Central Valley from mother lodes in the mountains.

When gold was discovered in 1848 on the south fork of the American River, a tributary of the Sacramento River, the news lured 100,000 men to the region. The cultivation of the valley's real wealth properly began when placer-gold supplies ran out within ten years of gold's being found. Would-be gold miners returned to their natural skills. As a consequence the Central Valley of California is now the market garden of North America.

A present-day model for the formation of parts of the Coast Ranges of California is shown here. (The sketch is based on a diagram provided by Warren Hamilton, of the USGS, Denver, Colorado.) The continental edge of southern

Sumatra is forming a mélange of rock by scraping along the ocean crust beyond the Java Trench. The angle and proportions of interacting surfaces are to approximate scale.

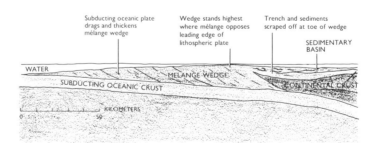

Subducting oceanic plate drags and thickens mélange wedge

Wedge stands highest where mélange opposes leading edge of lithospheric plate

Trench and sediments scraped off at toe of wedge

SEDIMENTARY BASIN

WATER

SUBDUCTING OCEANIC CRUST

MELANGE WEDGE

CONTINENTAL CRUST

KILOMETERS

0   50

between them and the Sierra Nevada batholith. The basement rock of the trough became the floor of an elongated sea-filled basin, in which sediments were deposited. The trough began to fill first with volcanic debris and later with alluvium eroded from the Sierra Nevada and to a much lesser extent from the Coast Ranges. A section through the great Central Valley of California today (for this is what the trough became) reflects the waxing and waning of magmatic events and the accumulation of the sedimentary products of erosion through a period of 135 million years.

San Francisco Bay is a sea-filled depression (an embayment) in the present-day Coast Ranges. The Golden Gate is the only break in an otherwise landlocked domain, the weakest link in the chain of coastal mountains through which tidal waters ebb and flow at a rate many times greater than the mean discharge of the Mississippi River. Tides, pounding by the sea, and the perpetual flow of meltwater from the accumulated snow and glaciers of the Sierra Nevada throughout the ages have combined to cut through passing suspect terrain as it slipped along the coast outside the Golden Gate. Some think that there may once have been a break similar to the Golden Gate in the vicinity of Monterey Bay, 100 kilometers south of San Francisco Bay. Whatever the original outlets of yesteryear, for human purposes it would not be possible to improve upon the present natural harbor and the great Central Valley of California that it serves.

The Central Valley is about 725 kilometers long and averages 90 kilometers in width. The length of its eastern side is dominated by the now sharply profiled Sierra Nevada, which contains literally hundreds of granite peaks, many over 4,000 meters. The valley's western edge is flanked by the low-profile Coast Ranges, which average a third of that height. The prevailing Pacific winds are semi-tropical and westerly. The Coast Ranges, for the most part ascending abruptly from the seashore, force the moisture-laden winds to rise and then precipitate moisture as fog, mist, and rain, which lowers air temperature along the coast. A temperate climate is the result, and so the coastal regions are lush and heavily timbered. The air temperature in the San Francisco Bay area is finely balanced throughout the

year and varies by only a few degrees through summer and winter. The bay area has an ancillary air-conditioning system, the result of the huge volumes of cool ocean water that twice a day flood into and out of the bay via the Golden Gate. In summertime warm moist air from the Pacific, lower early-morning temperatures, and consistently cool harbor sea water cause coastal fog to fill the bay before sunrise.

The long, broad, northwest-southeast Californian valley, the Central Valley, is a natural garden, with an encouraging variety of well-aerated light soils and an extraordinary climate. Although adjacent to the cool San Francisco Bay and Coast Ranges, the valley's fertile soils benefit from dependably warm, cloudless days with temperatures up to 40°C in summer in the south and a cooling gradient toward the north. The subtropical air that crosses above the valley from west to east is wrung dry by the Sierra Nevada peaks. For three seasons in each year there are extraordinarily heavy snowfalls in the mountains. To produce tributary streams, water from the snowfields in spring and early summer gradually discharges from south- and west-facing inclines into west-running valleys. The streams flow at right angles to the two main streams running down the center of the valley from its head and from its foot to its center: the Sacramento River from north to south and the San Joaquin River from south to north. The rivers meet near the center of the valley and discharge into San Francisco Bay. All factors combine to make the Central Valley the perfect agricultural machine.

In the 16th century Shakespeare would have been quite justified in referring to the "Vale of California" with its redwood coastal forests, its lush interior plain, its glaciated mountain valleys, its waterfowl-populated marshes, its salmon-filled rivers, its regulated climate, and its aboriginal inhabitants, in the same vibrant terms that in fact he coined to describe the England of his day:

> This other Eden, demy paradise,
> This Fortresse built by Nature for her selfe,
> Against infection, and the hand of warre:
> This happy breed of men, . . .

In the last reference Shakespeare had in mind Sir Francis Drake and his Elizabethan

contemporaries. Drake, like his successor, Captain James Cook, two centuries later, had been instructed to seek the Northwest Passage from the Pacific into the North Atlantic in his ship, the *Golden Hind*, along the "backside of Nueve Espanna" during his voyage of circumnavigation. According to Francis Fletcher, a "preacher" in Drake's employ on board the ship, whose notes (among others) were edited by Drake to produce an account of his voyage entitled *The World Encompassed* (1628), the *Golden Hind* sailed from Guatulco, Mexico, on April 16, 1579, to search for the Northwest Passage. Having reached as far north as the Oregon coast by June 3, where "ropes of our shippe were stiffe, and the rain which fell, was an unnaturall congealed and frozen substance . . .," they continued north as far as 48° (just south of Victoria Island, though it is considered by some to have been unlikely that the *Golden Hind* reached so high a latitude). After running into the "most vile, thicke, and stincking fogges; against which the sea prevailed nothing . . ." and suffering considerably from the cold, they made about to seek a "convenient and fit harbrough, and on June 17. came to anchor therein: where we continued till the 23. day of July following." The "harbrough" is believed to have been Drake's Bay, a part of that perfect example of suspect terrain called Point Reyes, 45 kilometers north of the Golden Gate.

Although the matter is contentious, it seems from Fletcher's account that Drake never did see San Francisco Bay; for Fletcher could hardly have failed to make special mention of such a harbor if Drake had discovered it. When referring to the ship's departure from California on July 23, 1579, the account continues: "Not farre without this harborough did lye certain Ilands (we called them the Ilands of Saint James) having on them plentifull and great store of Seals and Birds, with one of which we fell July 24. whereon we found such provision as might competently serve our turne for a while." If we assume the ship's departure point to be in fact Drake's Bay, the only sizable islands within easy reach are the Farallon Islands, 45 kilometers south-

This section cuts from west to east through one of the most interesting geological regions on Earth. It shows the geology of the North American continent from San Francisco Bay in California, across Nevada, to Salt Lake City in Utah and the Wasatch Mountains beyond, near the northwestern extremity of the Colorado Plateau.

The Sierra Nevada mountains were formed on the edge of the continent, which at that time was well to the east of its present position. About 135 million years ago the North American continent was being conveyed westward over the Farallon Plate at a time when the Atlantic Ocean was, in terms of geological time, rapidly widening during the breakup of Pangea. The Sierra Nevada was formed by

the interaction between the continental crust and the ocean plate's subduction beneath it. The basement rocks of the continental craton melted to form plutons, which streamed up toward the surface and accumulated to form a gigantic batholith.

The Coast Ranges of California were formed by the leading edge of the continent, scraping along the surface of the ocean crust as it subducted beneath the continental margin. The space between the Sierra and the Coast Ranges was once an open, sea-

filled basin, which gradually silted up with the products of erosion from the Sierra Nevada and volcanic debris. The basin was eventually filled except for an area of marsh and an embayment. Today the basin is the Central Valley of California, and the embayment, a depression in the Coast Ranges, is San Francisco Bay.

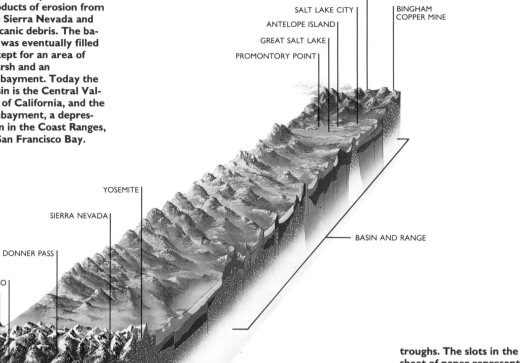

troughs. The slots in the sheet of paper represent weaknesses, or incipient fault lines, in the continental crust. It was through lines of such structural weakness in the continental crust that volcanic eruptions released vast quantities of lava onto the surfaces of Utah and Nevada during the stretching process caused by the subduction of the ocean plate. The vast area that was stretched and deformed is now called the Basin and Range Province and is some of the thinnest known continental crust on earth. It is still being actively stretched. The Basin and Range Province is the site of Death Valley, the Funeral Mountains, and the Great Salt Lake — among many other extraordinary geological features. (See the chapter entitled "Mobile Belt.")

During the formation of the Sierra Nevada huge amounts of relatively cold basaltic material from the subducting ocean crust plunged beneath the continental craton beneath and beyond the Sierra Nevada.

The subducting ocean plate caused the continental crust above it to lift and spread. The region was stretched to almost twice its original width as a result. The continental crust behaved rather like a sheet of paper with parallel slots cut into it. If both ends of the paper are gently pulled apart while the surface of the paper is pushed upward from beneath, the paper will expand into peaks and

SAN FRANCISCO BAY, CALIFORNIA

San Francisco Bay is one of the most spectacular natural harbors in the world. It is formed in a depression in the Coast Ranges. A shallow inland sea once filled the area of the Central Valley between the Sierra Nevada and the Coast Ranges of California. The inland sea silted up with volcanic debris and later with alluvium and glacial deposits carried down from the surrounding mountains. Two lateral river systems became established as the valley floor formed: the Sacramento River from the northern end of the valley and the San Joaquin River from the southern end. Both rivers have tributaries joining them at right angles from valleys that have cut back into the Sierra Nevada. Today the only gap in the Coast Ranges that provides an outlet for this freshwater system is between the headlands of the Golden Gate, which were linked by the Golden Gate Bridge in 1937.

The San Andreas Fault crosses the Farallones Gulf on the seabed behind the camera position. It enters the sea at Bolinas Lagoon to the north and emerges on land again near Daly City at the extreme right of the panorama. Perhaps one of the best known earthquake disasters in recorded history — but by no means the most severe — took place at 5:13 A.M. on April 18, 1906, when the mobile section of the San Andreas Fault suddenly moved 5 or 6 meters to the north. Much of the original city of San Francisco was wrecked by the violent tremors that resulted, and the center of the city was destroyed by the fire that followed. About 700 people died, and a quarter of a million were left homeless. It is a sobering thought that one inevitable day there will be another major lateral movement of the San Andreas Fault as the land west of the fault again moves northward. The population of the region has increased more than tenfold since 1906.

Drake's Bay, just south of Point Reyes and sheltered from the point's wild northwesterly winds, is thought to have been the place used as a "harbrough" by Francis Drake (later Sir Francis) for a period of nearly six weeks in the summer of 1579, during his circumnavigation of the globe. Contemporary notes edited by Drake and included in his book *The World Encompassed* (1628) almost certainly rule out San Francisco Bay, 45 kilometers south of Point Reyes, from his discoveries.

Contemptuous of Spanish claims to California, Drake in typical Elizabethan fashion claimed the region for England and called it Nova Albion ("New Britain"), supposedly because of the resemblance between the cliffs in the bay pictured above and those in the English Channel. Readers can make a direct comparison by referring to the photograph of the famed Seven Sisters at Beachy Head, England, on page 67.

DRAKE'S BAY, CALIFORNIA

west of Point Reyes in the Farallones Gulf. According to Fletcher's account, after an overnight stay Drake considered that the continued and deteriorating cold weather plus a strong northwestly wind were sufficient reason to abandon his attempt to tack northward and so "bente his course directly to runne with the Ilands of the Moluccas . . .," to which they continued "without sight of any land for the space of full 68, days together . . . ."

Francis Fletcher's detailed account of Drake's stay in California is absorbing. Point Reyes (if that was the place) and its neighborhood were populous in 1579: there were certainly hundreds of local inhabitants in a number of villages of earth-covered wigwam-like structures. The natives were armed with bows and arrows, the arrows in quivers made from deerskin; the natives were extremely adept in the use of these weapons, which at first made the Englishmen wary of too close contact. But as familiarity increased, relationships improved. According to Fletcher's account the inhabitants proved to be extremely athletic in many other respects. They habitually ran long distances and did so without tiring. They were strong, could carry on their backs as much as any two of the visitors, "and (without grudging) carrie it easily away, up hill and downe hill, an English mile together." After the ship's maintenance had been effected, Drake went ashore with his officers to visit the villages. They found the interior to be "far different from the shoare, a goodly Country, and fruitful soile stored with many blessings fit for the use of man: infinite was the company of very large and fat Deere, which there

we saw by thousands, as we supposed, in a heard . . . ." Despite their protests Drake and his companions were treated as gods, and to their consternation sacrifices were made in their name and in their presence, including severe self-inflicted injuries to faces and bodies, particularly among the women, whose duty it seemed to be to suffer in this way. The Englishmen were taken aback at the men's nakedness but seemed relieved that the women "take a kind of bulrushes, and kembing it after the manner of Hempe, make themselves thereof a loose garment, which being knit about their middles, hanges downe about their hippes, and so affords them a covering that, which Nature teaches them should be hidden: about their shoulders they weare also the skin of a Deere, with the hair upon it." After almost six weeks the parting between the athletic natives and their fur-coated women and the English sailors was apparently regretful. According to Fletcher's record, confirmed by Drake's editing, there seems to have been a happy first contact between the Europeans and the aboriginal Californians.

Recent estimates suggest that well in excess of 10,000 aboriginal people were in the bay region in the 16th century, and at least 350,000 in California as a whole. The bay people were not a cohesive group but lived in scattered villages as did their brethren in other areas of California. Those visited by Drake were probably typical. Each village is thought to have had up to 250 inhabitants, and each community lived independently of its neighbor, speaking different versions of a similar language. Life for the *ohlone*, the local collective term for the aboriginals in the bay region, was in many ways idyllic; for the natural bounty of Califor-

nia was unsurpassed. In comparison with life in Beringia or the far northwest Pacific coast or the deserts of the Southwest, the San Francisco Bay area was indeed "This other Eden."

The first undisputed time that the bay was seen by Europeans was in 1769, when a scouting party sent out by the Spanish explorer Gaspar de Portola saw it from a hilltop. It was not until six years later that a ship sailed into the bay, the *San Carlos* commanded by Juan Manuel de Ayala. The following year Spanish settlers from Monterey arrived on the southern peninsula overlooking the Golden Gate Narrows, and the San Francisco de Asís mission began ministrations the year after that. Half a century later, in 1835, so tradition has it, an otherwise unknown Englishman, Captain William Anthony Richardson, claimed a space for himself near the shore a few miles east of the mission and erected the first dwelling on the site of the future city of San Francisco. That dwelling was little more than a tent made from a ship's sail and a few pieces of redwood, but the site was attractive and slowly grew into the village of Yerba Buena. In 1846, when America was at war with Mexico, a man-of-war, the sloop *Portsmouth*, sailed undenied into the bay. The ship's captain, John B. Montgomery, promptly raised the American flag at Yerba Buena. Six months later, on January 30, 1847, the town (now of a thousand souls) was renamed San Francisco — the name just sounded more attractive. Almost a year to the day after this name change this blissful, unpretentious, promising, maritime

This nuclear power plant at Diablo Canyon, near San Luis Obispo, California, has been the subject of public protest against the dangers it might one day present. It was built on suspect terrain of the mobile San Andreas block. The next time that the block moves northwestward, the nuclear plant might survive the test and thus prove to be a triumph of man's ingenuity. The protestors fear that the plant could equally well prove to be a monument to man's hubris.

DIABLO CANYON, CALIFORNIA

and rural community became the focus of world attention.

On the morning of January 24, 1848, nine days before California and much of the Southwest were ceded to the Union by Mexico, pea-sized nuggets of gold were discovered in a newly deepened millrace at John A. Sutter's sawmill in the gentle Coloma Valley, through which ran the south fork of the American River—a westward-flowing tributary of the Sacramento River. One of the great secrets of the Sierra Nevada had been uncovered, and the mad scramble to benefit from the discovery began. In the next two to three years 100,000 gold seekers arrived in California by sea, by land from the east via Fort Laramie and the wagon trails, and from southern California and Mexico. By 1851 hundreds of ships rode empty at anchor in San Francisco Bay, abandoned by their crews who had joined the rush. Murder and mayhem was frequent. In 1856 the *Sacramento Union* commented that there had been "fourteen hundred murders in San Francisco in six years and only three of the murderers hung, and one of these was a friendless Mexican." Sutter's millrace and the ransacked banks of myriad neighboring streams, mostly ran out of gold in 1857.

By that time tens of thousands of new Californians in San Francisco were rich from selling eggs to the needy at a dollar each and other supplies and services to match inflated gold-rush values. Most diggers for gold found none and were bereft of financial means to return home. Many of them turned to what they knew best: the tilling of soil. Commerce in San Francisco and agriculture in the Vale of California were suddenly an established and necessary means of continued survival. Only a century later California had become the most populated state in the Union, with 4 million people clustered in cities around San Francisco Bay and 20 million in California as a whole. Commerce and agriculture have flourished, but the original 1,800 square kilometers of San Francisco Bay have been reduced by a third by

NAPA VALLEY, CALIFORNIA

landfill (now severely restricted), which has been turned by this means into valuable real estate. This reduction in water area has reduced the effectiveness of the delicate air-conditioning system of the bay, increased the effluent, and destroyed much of the wildlife.

Today it is the water supply of the Central Valley rather than its placer gold that inflames passion. Some regions in the northern sector of the valley have annual rainfalls of over 40 centimeters (Humboldt County). To the south beyond the Tehachapi Mountains, which enclose the end of Central Valley, the average rainfall is 3.8 centimeters annually. But 60 percent of California's population prefer to live in this dry desert-climate region, as witnessed by the growth of Los Angeles, San Diego, Palm Springs, and hundreds of satellite cities, towns, and villages in Southern California. Although these communities already have access by extraordinary aqueducts to the outflow of the Colorado River, which would otherwise drain into the Gulf of California, and to water drained from the eastern side of the Sierra Nevada, the need is greater than the available means. The California State Water Project launched in 1960 is designed to transfer 6,800 million liters daily from the Sacramento and the San Joaquin rivers near San Francisco, by a waterway 550 kilometers long, to Southern California: the largest civil-engineering project of its kind ever attempted. Central Valley agricultural and commercial interests fear for the future of their valley; Southern Californians insist on what they consider to be their rights. Tempers run high. The destiny of many people and the future of Central Valley could be at stake.

Here in the Napa Valley, nestling between volcanic hills with well-drained soils and an unmatchable climate, is one of the world's ultimate wine-growing areas. Once blossoms on the vines are set and fruit begins to grow, the rate of growth and degree of ripening, and therefore the quality of the wine that the grapes can produce, is controlled by weather. In the Napa Valley not only is the weather extraordinarily dependable, but there are also three distinct microclimates within the valley, which fine-tune the good effect of that dependability.

The summer mists from San Francisco Bay to the south of the Napa Valley percolate upvalley to the north. In the morning the mists slowly dissipate from north to south in the valley in the warmth of the sun. The mist forms again late in the evening as moist air sweeps in from the bay. The north end of the valley therefore has the highest number of hours of sunshine, the south end has the lowest, and the center of the valley has a gradient between the two extremes. This natural zoning of cool-air and hot-sun conditioning provides the grower an unusual degree of control over the development and ripening of particular varieties of grape. Such conditions, allied to new wine-making technology freed from the inhibitions of tradition, allow vintage-quality wines to be regularly achieved.

The ultimate compliment that can be paid by French wine makers to growers in California is to agree to the production of a joint vintage. One of France's leading vintners, Philippe de Rothschild, of Château Mouton Rothschild, Pauillac, has joined with Robert Mondavi, a leading Napa Valley vintner, to produce the first Franco-American Californian vintage. This was bottled in 1979 and will be released late in 1983. The first twenty cases of the wine, a Cabernet Sauvignon, were sold at a wine auction for $2,000 a case, the highest price paid for an American wine and a record for any "future" wine.

POINT LOBOS, CALIFORNIA

There are many blissful places on Earth where ocean and sky meet continental shores, but there can be few meetings that surpass Point Lobos near the Monterey Peninsula in California. The natural scene beside an emerald sea is set in early-morning mist on rough, rocky cliffs slapped by Pacific rollers. The cliffs are surmounted by wild, gesticulating stands of dark and threatening Monterey cypress. The secret sheltered places are carpeted with flowering succulents tumbling in clusters almost down to the sea itself. As the sun burns through the mist, colors change and edges sharpen. The ocean turns an azure blue, and the sea spray upon the cliffs to a scintillating white. Kelp-filled coves are suddenly populated with female sea otter feeding their young while floating on their backs, the males perhaps busily pounding abalone on a rock balanced on their chests. Pelicans fish, and cormorants dive. A noisy chorus of sea lions offshore plays counterpoint to the low rumble of waves. And one man's small world is momentarily at peace.

John Muir, born in Dunbar, Scotland, in 1838, immigrated with his parents to Portage, Wisconsin, in 1849. Muir became the first man publicly to recognize the possible dangers of excessive human influence on the California "demy paradise." By profession Muir was an engineer, but by inclination he was a naturalist. After walking from the Middle West to the Gulf of Mexico in 1867 and the next year discovering for himself the glorious glaciated granite valley of Yosemite in the Sierra Nevada, this dour and far-seeing Scot devoted his life to the natural history of California and the cause of conservation. He saw in forthright terms the dangers of indiscriminate felling of giant trees, the coastal redwoods and the mountain sequoias, many of which need several thousand years to grow to their present size. He anticipated the destruction of what he recognized to be the delicate ecological balance on which the magic of Yosemite Valley depends. Muir's weapons in his defense of these and other places were his intellect and his considerable powers of public and political persuasion. Mainly as a consequence of his efforts, the Sequoia and Yosemite na-

tional parks were established in 1897. Later he greatly influenced the establishment of the U.S. national conservation program, initiated in 1903. His most long-standing contribution to a sometimes reluctant human society was the founding of the Sierra Club in 1892. The Sierra Club today maintains its founder's tradition of focusing public outrage and political influence in the interests of conservation.

There are many distinguished members of the Sierra Club. They are pilloried by some and revered by others. To the majority of the public, however, the distinguished members have become formidable guardians of the wilderness. But in fact the Sierra Club, like all else in California, is a by-product of past subduction of the Farallon Plate beneath the North American Plate.

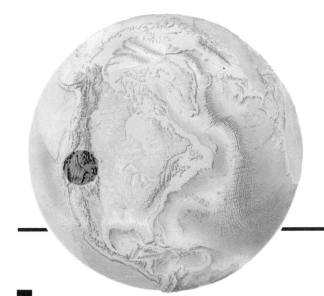

MOBILE BELT

IT was just one of those days. An old friend, Martin Litton, had joined Joy and me at the Madonna Inn at San Luis Obispo, California. Joy and I had driven down there along the Big Sur coast road from Point Lobos after flying up and down the San Andreas Fault and the California coast with Martin for a week. We had determined to make a predawn start the following day to do some aerial photography over the Carrizo Plain, Mount Whitney, Owens Valley, and Death Valley, the first leg of a long trip that would conclude at Yellowstone, Wyoming.

In the predawn half-light I had packed most of the luggage in the rear of the frost-rimed rental car, slammed the lid, and put the rest of the gear on the back seat preparatory to moving off. Only then did I discover that the one and only car key, which had been in my hand moments before, was now securely locked inside. After what seemed an interminable wait two burly locksmiths arrived. One of them gained access in two seconds flat. We drove to the airport and returned the car. Then Martin tried to start his beloved Cessna 195 — the ideal plane for photography, with its high, strutless wings and powerful radial engine (and an oversized variable-pitch propeller, which Martin had added). It had been a cold night: The engine wasn't interested. A fraction too much choke, a flood in an overheated carburetor, and we had a nice fire going inside the engine cowling. Martin promptly put the fire out with a carbon dioxide extinguisher. The engine was now warm. We tried again, and it coughed into reluctant life. We took off for the Carrizo Plain and the most spectacular part of the San Andreas Fault, which crosses it; the fault and the

Temblor Range beside it were shrouded in fog. There was to be no low-lit early morning photography. It was one of those days all right. But from an altitude of 3,500 meters we could see the sunlit peaks of the Sierra Nevada on the horizon, 300 kilometers to the northeast, and the highest peak, Mount Whitney (4,400 meters), was free of cloud.

Litton is a burly Californian, an ebullient ex-journalist with a fierce mop of silver-gray hair and a generous moustache to match. To see him row one of his Grand Canyon dories through a really big Colorado River rapid in fast, low, and dangerous water — a time when passengers just stand with bated breath and watch from the safety of the river bank — is to know how he is going to fly an airplane — with exacting flair and exciting panache. I flew with many pilots during the course of photographing for this book, but none could

FROM MOUNT WHITNEY TOWARD OWENS VALLEY, CALIFORNIA

match this man's skill, not even the retired jumbo-jet pilot in the Shenandoah Valley who later flew Joy and me in a little Cessna 172 as if it were a Boeing 747. Martin invariably put his plane where the photographer needed to be, at exactly the right height and angle and frequently after a complicated preliminary maneuver accomplished in difficult flying conditions in dangerous places; the comparison with running Grand Canyon rapids is apt. The first leg of the route lay along a northeasterly line from the nuclear plant at Diablo Canyon, via the San Andreas Fault at Carrizo, to Death Valley, the Funeral Mountains, and a Nevada testing site for nuclear bombs. On reflection this seems not to have been too tact-

ful a line to have chosen; for Martin Litton is a fervent environmentalist and one of the leading lights of the Sierra Club.

When one has learned how the Sierra Nevada mountains were formed, they take on a different significance. It's rather like learning something both extraordinary and unsuspected about an old acquaintance; one's perception changes, and that person never seems quite the same again. For me Half Dome and El Capitan and other spectacular formations in Yosemite Valley immortalized in words by John Muir, on canvas by Thomas Moran, and on film by Ansel Adams are no longer *just* beautiful. Their form, color, and majesty have been marvelously enhanced. When I looked down as we approached Mount Whitney from the west, I appreciated that I was not looking at separate mountains. They were a cohesive

whole, a multiplicity of plutons, a batholith that had been eroded into spikes and towers and flat-topped skyscrapers like Mount Whitney itself. The Sierra Nevada is a magmatic agglomeration interspersed with Alpine meadows, lakes, and streams that had nurtured the giant sequoia trees.

But beyond Mount Whitney and the eastern escarpment of the Sierra Nevada there is an abrupt end to tranquility. Suddenly the world of snow-covered peaks and gentle valleys becomes barren and bleak, a world struggling to maintain a stark but cruel beauty in the rain shadow of the batholith. Much of the Sierra escarpment is a sharply drawn geological boundary between lush meadows and dry desert. The surface of the craton in Owens

This is the view due east from Mount Whitney (4,400 meters), California. The basin immediately below the Sierra Nevada is Owens Valley. The base of Mount Whitney and the edge of Owens Valley meet at the boundary of the Basin and Range Province at its narrowest point. The Inyo Mountains form the horizon, and there is an alternation of basins and ranges for 400 kilometers beyond. Death Valley, the Funeral Mountains, and Las Vegas lie almost directly between this point and Grand Wash Cliffs at the end of Grand Canyon, which marks the other boundary of this section of the Basin and Range.

The province stretches 2,000 kilometers from Idaho in the north to Sonora, Mexico, in the south and sweeps eastward around the Colorado Plateau to southern Arizona and New Mexico. This region is sometimes called the "Great Basin" and is one of the largest deserts on Earth.

The continental crust below the Great Basin is about half its usual thickness because this part of the North American continent has been stretched to twice its original width. Occasional severe earthquakes and volcanic eruptions in Owens Valley and at Mono Lake out of the picture to the left are manifestations of continued stretching.

Valley at the base of Mount Whitney extends from Oregon and Idaho in the north to Sonora, Mexico, in the south, and to the edges of the Wasatch Mountains, the Colorado Plateau, southern Arizona, and New Mexico in the east and southeast. This huge area is one of Earth's great deserts. It is in excess of 2,000 kilometers in length and varies from 300 to 800 kilometers in width. Most of the region was once a cratonic stable platform and is now part of an active mobile belt. It is a very hot and very dry region of great geological interest called the Basin and Range Province.

When the early sedimentary rocks of the Grand Canyon were being formed in pre-Pangean times, the western margin of the continent was passive. The relationship was similar to the present relationship between the East Coast of North America and the Atlantic Ocean crust. But as North America collided with Baltica (now Northern Europe) to form the Appalachian Mountains on the eastern edge of the continent, so the ocean crust along the western extremity began to subduct. Subduction continued until Pangea had fully formed, was intensified when Pangea began to split up, and vigorously increased when the Atlantic Ocean began to open.

The rocks exposed in Grand Canyon are representative of the rocks that once extended to the Pacific Ocean. They epitomize the character of the craton at the continental margin to the west, which came under the onslaught of plutonic magmatism during the interaction between the descending ocean plate and the advancing continental crust. The Sierra Nevada plutons penetrated the Precambrian Shield and stable platform above. But the Sierra batholith was just one of a chain of batholiths along the whole western length of the North American continent. Each batholith was formed by different parts of the ocean plate that interacted in different ways with the edge of the continent. There is the Baja batholith to the south of the Sierra batholith, and there are other batholiths to the south of Baja. To the north of the Sierra batholith there is the Idaho batholith, and there are others north of Idaho, including the vast British Columbian batholith.

This confrontation between continental and ocean crusts caused the formation of a mountain chain along the Pacific coast that was of

This paleogeographic map depicts the assembly of continents prior to the formation of Pangea. About 400 million years ago the future North American continent (shaded area) was colliding with continents to the east of it.

the same stature and character as the Andes of South America today. Such mountains, or groups of mountains related to particular batholiths, are termed magmatic arcs (an island arc forms only on the ocean bed). In addition to such continental magmatic arcs forming on the edge of the continent, accretion of suspect terrains had taken place along the western margin before Pangea was formed. Further accretion occurred during the formation of Pangea, and even more accretion of microcontinental terrain, chunks of displaced ocean bed, and ocean seamounts took place during the post-Pangean episode (as described in the previous chapter).

The ocean crust that was subducting beneath California up to about 30 million years ago was associated with a spreading ridge similar to the Mid-Atlantic Ridge. This spreading ridge is called the East Pacific Rise. The ocean crust that once subducted beneath California from the west (and therefore to the east of the East Pacific Rise) is called the Farallon Plate. The ocean crust to the west of the East Pacific Rise is the Pacific Plate, which, because of the almost total destruction of the Farallon Plate by subduction beneath California, is now adjacent to the continental margin along the San Andreas Fault. Even though comparatively little of the Farallon Plate remains today, it was its subduction that simultaneously caused the formation of three gigantic batholiths, the Baja, the Sierra, and the Idaho.

The only way that one can begin to appreciate the staggering scale of these events is to relate them figuratively to the present width of ocean crust between the Mid-Atlantic Ridge and the East Coast of North America (see the globe on page viii). If one imagines an ocean plate of this approximate size, 5.5 kilometers thick (roughly the original area and thickness of the Farallon Plate), to have been pushed beneath the western edge of North America, one begins to understand the scale of the events that occurred on the West Coast. In addition to the formation of batholiths on the edge of the continent subsidiary effects were caused by the enormous volume of basaltic material that descended beneath the continental lithosphere. One consequence of this subduction was the formation of the Basin and Range Province by a process called "back-arc

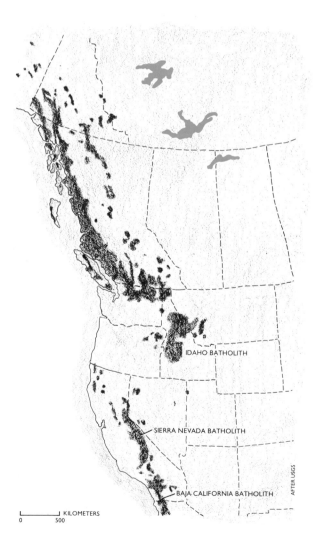

For a period lasting about 100 million years, which began around 180 million years ago, the Pacific West Coast was the scene of traumatic change as ocean crust subducted beneath the continent. Subduction caused a magmatic arc of batholiths to form on the surface above the zone of subduction. The resulting mountains were of Andean proportions. The Idaho, Sierra, and Baja batholiths were part of the same arc when they were formed, but they were later separated along the line of the Snake River Plain between Idaho and Utah and Nevada.

spreading." This occurred in two stages.

As previously described, the Sierra Nevada magmatic arc first began to form a batholith. Then the Basin and Range Province stretched, or "extended," behind it. The effect of such a geological extension of continental crust is analogous to the effect of stretching a flat sheet of paper, after first slitting it through in parallel lines, and then pulling at the edges of the sheet. The pulling causes the sheet to extend; and in so doing, the slits open up into slots that arch upward and twist. The surface of the extended sheet will now look like the Basin and Range, which has parallel block-faulted mountains running along its geographical length that correspond to the twisted slots. (See the illustration on page 113.)

The Basin and Range was being formed while subduction progressed beneath the continental margin. The farther the continent advanced to the west and the subducting Farallon Plate advanced to the east, the closer the continental margin came to the East Pacific Rise spreading ridge. Some geologists think that the continental margin passed over the spreading center and that rifting continued beneath the continental crust, thus contributing to the stretching of the Basin and Range Province. Another school of thought, apparently the majority view today, suggests that, when the continent met the spreading ridge, approximately 30 million years ago, the spreading zone simply ceased to function — but with important subsidiary effects.

The majority theory suggests that by 30 million years ago a point had been reached where the Pacific Plate came into contact with the North American Plate after a section of the Farallon Plate had been consumed beneath the continent. The Pacific Plate did not subduct on contact with the North American Plate but continued its original northwesterly movement, causing a fault to form on the continental margin: the San Andreas Fault. There had been an apparent change in plate motion relative to the western edge of the continent. But the real change was a change in plate relationship with the continental margin. The change was from the southeasterly moving and subducting Farallon Plate *east* of the spreading

ridge to the northwesterly moving Pacific Plate *west* of the spreading ridge.

Following the formation of the San Andreas Fault, the theory concludes, the East Pacific Rise interacted with the Mexican section of the continental margin, causing part of it, the "San Andreas block," to break away to form the Gulf of California. The San Andreas block, which includes Baja California and which had thus been split from the mainland by the San Andreas transform fault, began to raft up the Californian coast on the northwesterly trending Pacific Plate — and it continues to do this today. Similarly, the Basin and Range Province is also active; it is *still* being extended. At the northern end of the province a remnant of the Farallon Plate is still subducting beneath the continental margin, as evidenced by the recent eruption of Mount St.

80-60 MILLION YEARS AGO

PACIFIC PLATE

CONTINENTAL MOVEMENT

FARALLON PLATE

EAST PACIFIC RISE

30 MILLION YEARS AGO

EAST PACIFIC RISE

25 MILLION YEARS AGO

PACIFIC PLATE

EAST PACIFIC RISE

3 MILLION YEARS AGO

REMNANT FARALLON PLATE (JUAN DE FUCA PLATE)

MOUNT ST. HELENS

SAN ANDREAS FAULT

BAJA CALIFORNIA

EAST PACIFIC RISE

AFTER GEOLOGICAL EVOLUTION OF NORTH AMERICA

Volcanics

Subduction Zone

In the post-Pangean times the westward-moving North American Plate in the region of California was overriding a part of the Pacific ocean crust called the Farallon Plate. This reconstruction illustrates four possible stages during this event.

Between 80 and 60 million years ago the Farallon Plate was subducting at an oblique angle beneath California. By 30 million years ago most of the plate had been subducted, and the edge of the continent came into contact with the East Pacific Rise (the Pacific Ocean equivalent of the Mid-Atlantic Ridge). By 25 million years ago part of California had overridden the spreading ridge; and by geologically recent times most of the Farallon Plate had been subducted.

The main consequences of these events were that the relationship between the Californian edge of the continent and the ocean bed changed from oblique subduction to a transform movement: The subducted plate contributed to the formation of the Basin and Range Province; the remaining parts of the Farallon Plate east of the spreading ridge are still being subducted beneath the continent north of California, as indicated by the formation of the Cascade Range and the eruption of Mount St. Helens.

EUREKA VALLEY, CALIFORNIA

ZABRISKIE POINT, CALIFORNIA

These sand dunes in Eureka Valley northwest of Death Valley, photographed from a height of 3,000 meters above the ground, are a reversing formation; very high dunes form along the spine, which geologists call a "Chinese Wall." The dune system is on the move most of the time, but the net effect of prevailing winds, countered by winds from the opposite direction, is that the dunes change shape but not location.

Helens, and this subduction contributes to the extension of northern Nevada, Oregon, and Idaho. In the province's southern part, in the region of Owens Valley and Death Valley, the effect of fragments of the former Farallon Plate beneath the crust plus the effect of movements along the San Andreas Fault are contributing to continued extension. A main factor that indirectly controls the extension of the Basin and Range and therefore the movement of the San Andreas block is the spreading speed and the direction of spreading at the Mid-Atlantic Ridge. The widening of the Atlantic Ocean significantly influences the behavior of the North American continent relative to the Pacific Plate. Things haven't changed in this sense since the original breakup of Pangea.

Owens Valley runs north and south imme-

This panorama of Zabriskie Point near Furnace Creek in Death Valley illustrates the effect of water erosion on colorful, unconsolidated mudstone, which was deposited in a lake that once filled the valley.

When changes occur in arid climates, they can be sudden and complete. Such changes are often caused by flash floods roaring down gullies after a heavy rainstorm in the mountains. Over thousands of years such flash floods have caused the accumulation of the huge alluvial fans along the shorelines that are a dominant feature of Death Valley topography.

diately below the Sierra Nevada's Mount Whitney, and the Inyo Mountains rise steeply along its eastern edge. This basin-and-range sequence typifies the north-south-trending topography of the desert and mountains of the province as a whole. When one flies in a direct line east from Mount Whitney across Owens Valley and beyond the Inyo Mountains, one passes over Eureka Valley, the Panamint Range, Death Valley followed by the Funeral Mountains, and so on until one reaches Grand Wash Cliffs, which mark the very edge of the Colorado Plateau at the end of Grand Canyon, 400 kilometers from Mount Whitney.

Death Valley is a focal point in this narrowest part of the Great Basin, but it is not so much a valley as a "bottomless" trough. It is in fact a "sink." The floor of Death Valley is a dried-up lake bed that is sinking into its slot faster than it can now be replenished with sediments. The surface of its old lake deposits, which must be of considerable thickness, is now near or actually below sea level. By the time prevailing winds reach this sink, they are usually so dry that it rarely rains, and even if a small amount precipitates, it simply evaporates as it falls. But if by some freak weather pattern a moisture-laden air system passes low over the Panamints or the Funeral Mountains, the most devastating flash floods can result in Death Valley. Other than these rare instances Death Valley is extremely arid and hot; up to 50°C on a summer's day is not unusual. Because Death Valley is a sink, water from neighboring areas gravitates toward the valley, and for this reason the water table beneath the valley provides a more than adequate supply to the oasis at Furnace Creek. But Death Valley is by no means unique. It simply has the lowest elevation, the most arid climate, and the highest temperatures of a great number of dried-up lake beds in the Basin and Range Province.

We took off from Death Valley in Litton's Cessna 195 after an overnight stay at the Furnace Creek Ranch. The Ranch, complete with a well-watered golf course surrounded by palm trees, is only a short drive from the low-

BINGHAM COPPER MINE, UTAH

Low-grade coppper ore extracted from open-pit mines in the Basin and Range Province has been the source of most of the copper produced in the United States. The ore was formed by the hydrothermal intrusion of huge granite stocks, which are several kilometers in diameter. The stocks were formed as a result of magmatism associated with the stretching of the continent and the thinning of the crust throughout the province.

Bingham Canyon, which is pictured here, is an open-pit copper mine near the shore of the Great Salt Lake, Utah. It is one of the largest holes ever to have been dug into the upper crust. But the percentage of copper present in the ore has declined as removal from the mountain advanced.

A new method of microbiological "mining" is now being used at Bingham to extract the very low concentrations of copper previously discarded in millions of tons of waste-dump material. Acidified water is poured on top of the dumps, which encourages the growth of certain bacteria that in turn accelerate the leaching of

copper compounds. The action is similar to that which occurs in a garden compost heap.

Solar energy alone is used to extract salts from the Great Salt Lake. As shown in the smaller picture, huge areas of the shallow lake are simply cordoned off, and the already very saline water is allowed to evaporate in the hot dry air. The ponds are replenished during evaporation so that each pan is filled with salt at the end of a cycle.

GREAT SALT LAKE PANS, UTAH

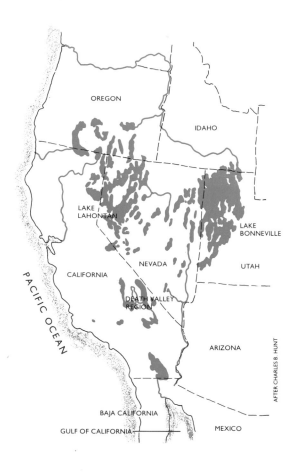

OREGON

IDAHO

LAKE
LAHONTAN

LAKE
BONNEVILLE

NEVADA

UTAH

CALIFORNIA

DEATH VALLEY
REGION

PACIFIC OCEAN

ARIZONA

AFTER CHARLES B. HUNT

BAJA CALIFORNIA

GULF OF CALIFORNIA

MEXICO

At different times during the last 2.0 million years the basins in the Basin and Range Province have contained lakes of considerable size. At the peak of the Pleistocene Ice Age they covered an expanse equal to at least half the present total area of the Great Lakes.

est point in North America (90 meters below sea level). We headed for Salt Lake City, 1,000 kilometers north-northeast. Some hours later as we neared the city, we could see that Salt Lake and its satellite towns have been built on alluvial fans accumulated at the foot of mountain-valley stream outlets on a wide bench between the original lake shoreline (Lake Bonneville) and what is called the Provo shoreline, about 100 meters lower. The present level of the Great Salt Lake of Utah is a further 200 meters below the Provo shoreline; the present lake is but a shadow of its former self.

During the different stages of the last Ice Age that ended its final major period of glacial resurgence (about 10,000 years ago) the character of the Basin and Range Province landscape was very different from that of today. Death Valley, for example, was at times one of a series of sometimes interconnected lake basins between the mountain ranges, and its overspill contributed to the Colorado River catchment area. But north of Death Valley these Pleistocene lakes covered vast and often totally enclosed areas. Water from the melting snows and glaciers of the Sierra Nevada to the west and from the Wasatch Range above the site of Salt Lake City to the east accumulated in the basins, leaving block-faulted mountain tops as islands in a freshwater sea. These landlocked lakes together covered an area equivalent to more than half the present area of all the Great Lakes. The largest single lake to accumulate in the Great Basin at that time was Lake Bonneville, estimated to have been about 50,000 square kilometers in area, roughly a third of the present size of Lake Michigan. The present 4,000 square kilometers of the Great Salt Lake of Utah are all that remain of Lake Bonneville.

The Great Salt Lake averages about 3 meters in depth, and this depth fluctuates considerably, depending on the volume of water flowing into it and the counterbalancing effect of evaporation. In contrast to the Great Salt Lake, Lake Bonneville at its peak level was over 300 meters in depth and would of course have completely submerged Salt Lake City and neighboring towns. Bonneville, however, was quite suddenly reduced, by about a third of its volume, to the Provo shoreline level during an event that occurred about 15,000 years ago. Harold E. Malde, a U.S. Geologi-

cal Survey (USGS) research geologist, completed a study of this remarkable event in 1968. In anticipation of my visit to the area I went to see Malde in his office in Denver and got from him a fascinating account of the Bonneville catastrophe together with the latest chronological information about the event.

Malde's work had shown that a breach of the shore of Lake Bonneville had occurred at Red Rock Canyon on the summit of a pass between present-day Utah and Idaho to the north of Salt Lake City. The breach was critical, causing Lake Bonneville to be lowered by 100 meters in a period of about six weeks. During this time the discharge rate at the breach averaged between 700,000 and 935,000 cubic meters of water per second. This is the equivalent of six times the present average flow at the mouth of the Amazon River, but the speed of the water from the breached lake was more than ten times the velocity of the Amazon.

Normally the lake did not accumulate sufficient water to overflow; evaporation prevented such excess. But at some point in the lake's history, perhaps during or after the capture of the Bear River (a previous tributary of the Snake River) by the Lake Bonneville basin, extra water accumulated to an unprecedented level. This caused an overflow stream to form at Red Rock Canyon. The overflow stream is believed to have cut back into retaining alluvial fan lying on a rock sill. Eventually a hairtrigger situation was reached. A meter or so of extra cutback into the unconsolidated alluvium, perhaps a collapsed rock or two downstream, and the overflow of water became a torrent and the torrent a catastrophic flood.

As lake water gained momentum through the spillway, it opened a gap that eventually became a breach some 3 kilometers wide. The enormous volume of water that then poured through the breach caused absolute havoc in the downstream basalt canyons of the Snake River Plain to the north and northwest of Red Rock Canyon. New channels and spillways cut into the basalt surface of the Snake River Plain measure 30 kilometers long, 300 meters wide, and 100 meters deep. In some places — as much as 300 kilometers below the breach —

where lava formations were softer the flood cut new channels 200 meters deep. Huge chunks of lava were ripped away from the surface of the plain and from the walls of the original canyons and then were rolled along by the floodwaters to form what has been termed "melon-gravel" bars. Some of these melons were almost house size, and thousands of them accumulated to form chaotic piles on the river bed when the water velocity dropped. Near the original Snake River Canyon the water cut a channel through the center of a volcano.

Once the level of Lake Bonneville had dropped to the point where there was consolidated rock at Red Rock Canyon that could withstand the flow of water, the flood subsided and the lake assumed a new shoreline, the Provo shoreline. The lake is thought to have remained at this level until about 13,500 years ago, when the climate began to change, heralding the end of the Wisconsin glacial advance. By about 10,000 years ago the lake had largely evaporated. A similar rate of evaporation reduced water levels in other lakes in the Basin and Range Province during this period. In due course most of them were evaporated entirely, leaving deposits of mineral salts.

We took off from a small airport near Salt Lake City, and there was little to remind us of the Hawaiian Islands as we crossed over Red Rock Canyon and completed our westerly flight down the bleak, uniformly colored Bonneville flood path on the Snake River Plain. But there is in fact a strong tectonic affinity between the plain, a 600-kilometer-long, 80- to 100-kilometer-wide, curved volcanic scabland (flood scarred), and the romantic islands in the middle of the Pacific Ocean. The affinity arises from the fact that the oceanic islands and the continental plain were formed by hot spots, or mantle plumes, beneath the upper crust. As the Earth's lithosphere moves over hot spots in the mantle, magmatic vents are burned through the crust. These magmatic holes leave a trail of volcanoes on the surface of the moving lithosphere.

In addition to their mantle-plume origin the Hawaiian Islands and the Snake River Plain have two further important common factors. Both have banana-shaped configurations,

which are indicative of the movement of the crust above them, and both have active volcanic regions at the "stem" of their banana shapes. In the case of the Hawaiian Islands the stem is presently Mauna Loa and Kilauea on the island of Hawaii although a young seamount on the ocean bed to its southeast, Loihi, is showing signs of activity and may soon form the youngest island in the group. The stem of the Snake River Plain is the Yellowstone caldera, better known as Yellowstone National Park, Wyoming.

We refueled at Boise, Idaho, near the terminus of the Snake River scabland, and flew due east toward Jackson Hole, which has the nearest large airport to Yellowstone National Park. Jackson Hole is also the locality of the quite spectacular Grand Teton Range, which, like Yellowstone, has an intimate but entirely different connection with the Snake River Plain. We cut across the southern foot of the Idaho batholith, and as we did so, I was reminded that the Idaho, Sierra, and Baja batholiths were formed simultaneously along the Pacific coast as a consequence of the Farallon Plate's subducting beneath the continental crust but are now separated laterally by a

distance of about 300 kilometers. The Snake River Plain was very much involved in this separation.

When the Farallon Plate had been consumed beneath California and the San Andreas Fault was established, only the northern part of the Farallon Plate remained east of the spreading ridge, and this continued to subduct beneath the Pacific coast in the region of the states of Oregon and Washington. This fragment of the original Farallon Plate (the Juan de Fuca Plate) caused the formation of a magmatic arc (the Cascade Range, which includes Mount St. Helens) and the volcanic flood plains of the Pacific Northwest. The extension of the Basin and Range Province had the effect of pushing the Sierra Nevada westward at a greater velocity than that of the general westward movement of the continent. The interaction of the Juan de Fuca Plate's resisting the advance of the continent to the west with the Basin and Range Province extension's encouraging its advance caused an offset of the Sierra and Idaho batholiths along a fault line represented by the Snake River Plain.

The Snake River Plain is a feature of tremendous interest to tectonic scientists. It is not known how or why the mantle plume that was once beneath it formed — only that the

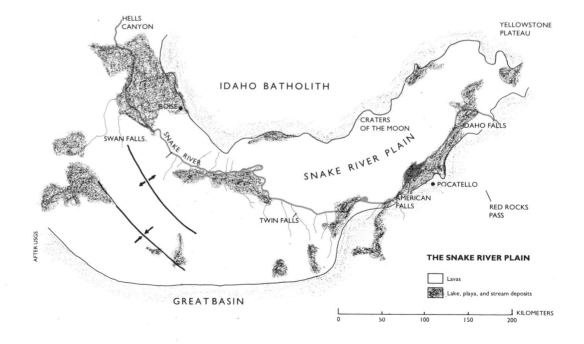

THE SNAKE RIVER PLAIN

☐ Lavas

▨ Lake, playa, and stream deposits

0   50   100   150   200 KILOMETERS

RED ROCK CANYON

ANTELOPE ISLAND, GREAT SALT LAKE, UTAH

SWAN FALLS

TWIN FALLS

DIVIDED VOLCANO

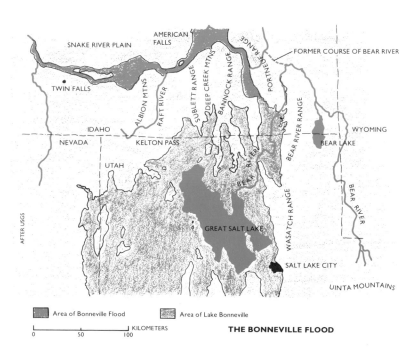

Area of Bonneville Flood   Area of Lake Bonneville

KILOMETERS
0   50   100

THE BONNEVILLE FLOOD

AFTER USGS

The largest of the Pleistocene lakes in this region was Lake Bonneville. It was more than ten times the area of the present Great Salt Lake, which is all that remains of it today. The shoreline of Lake Bonneville, about 300 meters above the present lake level, is marked by the highest bench that can be seen traversing the upper slopes of Antelope Island in Great Salt Lake (photographed here).

About 15,000 years ago Lake Bonneville was reduced by about a third of its volume during a cata-strophic breach at Red Rock Canyon (picture above, left) on the present Idaho-Utah state line. The level of the lake after the catastrophe is marked by the Provo shoreline, which traverses Antelope Island about 100 meters below the Bonneville bench.

The Bonneville flood lasted about six weeks. The volume of discharge at Red Rock Canyon was the equivalent of six times the present volume of discharge of the Amazon River. This enormous flow of water devastated the Snake River Plain in Idaho. An idea of the power of the flood can be gauged from the size of the generating plants at Swan Falls (picture at the left) and at Twin Falls, which is pictured above, center. In both instances the surrounding Snake River canyons were filled to their rims and beyond with water that reached a maximum velocity of about 70 kilometers per hour. The last small picture shows a volcano (top and left center) that was actually cut in two by the flood.

GRAND TETONS, WYOMING

The Teton Range of mountains is formed from a huge block of continental crust that has been uplifted and tilted. Some geologists believe that this block may be sliding into the tectonic hole represented by the Snake River Plain to the west and with the Yellowstone caldera at its head. In this panorama one can see the front of the uplifted and tilted block. The Snake River runs from right to left in the picture, passes around the block at its southern end, and then flows down the Snake River Plain beyond.

Grand Prismatic Spring (right) near the Firehole River in Yellowstone National Park is the epitome of Charles Darwin's "warm little pond," in which he visualized life to have begun.

hot spot still exists at the head of the plain. Warren Hamilton, of the USGS, Denver, during a question-and-answer session to which he kindly submitted, suggested that the Snake River Plain may be a rift in the continental crust with a magmatic feature at its point. In other words there is a hole through the continental crust there — not simply thick layers of volcanic rock piled up on existing rock but a lava-filled hole that may be necking down through the lithosphere into the asthenosphere. Research recently completed suggests that there have been a series of Yellowstones that formed and became extinct in a successive string, just like the Hawaiian Islands. There is, for example, another now completely extinct caldera to the west of Yellowstone, the Island Park caldera, which formed about 1 million years ago. At present the mantle plume that caused the Island Park caldera (and, presumably, the rift as a whole) is below Yellowstone National Park. It is Yellowstone that is over the hot spot today. Depending on the vagaries of the Mid-Atlantic Ridge, which controls the speed and direction of the continent's movement, the region now northeast of Yellowstone should be above the hot spot at some time during the next half-million years.

Hamilton went on to suggest that at the top of the Snake River Plain (immediately east of Yellowstone) there are major and very active young fault blocks, of which two are particularly conspicuous: the Teton Range on the southeast of the Snake River Plain and the Centennial Range on the north side of the plain. These features are quite symmetrical, and both are pointing straight at Yellowstone. One possible reason for both sets of young fault-block mountains is that they are tending to slide into the tectonic hole represented by the Snake River Plain. The valleys in front of both ranges, Jackson Hole and Jackson Lake in front of the Tetons and the source of the Missouri River in front of the Centennials, might be gaps caused by the pulling of the blocks as they part from the continental surface in their tendency to slide into the hole. But, Hamilton warned with a sudden smile (as he had once warned me before), "other people will have other explanations: You must make the choice."

Yellowstone is an enormous caldera formed by mega-explosions. The material that normally forms a volcano — the ash, the pumice, and the breccia — was so violently distributed that traces of it are found for many hundreds of kilometers around. When the eruptions from the magma chambers stopped, the region collapsed into a caldera, a thin and slowly inflating crust now lying querulously upon the surface. Because of its open position at the top end of the vast Snake River Plain to the west,

the Yellowstone caldera receives a very heavy winter snowfall. Spring, summer, and autumn in the region are crammed into a few short but very beautiful months — May to September. But the climate and the altitude contribute mainly to the accumulation of snow and water both above and beneath the porous volcanic surface. That there is a superabundance of water in an area simmering above a hot spot contributes to Yellowstone's main attraction to visitors, the renowned geyser and hot-spring basins. The whole caldera was in fact the first national park to be designated in the U.S. (1872).

Geysers and hot springs are caused by water that percolates downward from basins above the now-quiescent and occasionally grumbling magma chamber. The basins usually form in regions of the Yellowstone caldera that contain friable rock above solidified lava. The water penetrates to the region of this lava and is heated to temperatures in excess of the normal atmospheric-pressure boiling point. When the already high temperature of such water is raised above a critical point, the flash point, it converts explosively into high-pressure steam. A chain reaction results in which the initial explosion ejects the water column above it, which also converts to steam as pressure upon it lessens as it nears the surface. The

GRAND PRISMATIC SPRING, YELLOWSTONE NATIONAL PARK, WYOMING

The saguaro, the central feature in the picture to the left, is the dominant cactus species of the Sonoran Desert in the southern Arizona segment of the Great Basin. Thousands of these tree-sized plants create a weird scene to the unaccustomed eye: like a forest of trees stripped of most branches and all their leaves. In the left foreground of the picture is the teddy-bear, or cholla, cactus and in the right center an ocotillo, a deciduous plant with spikes and a cap of bright red flowers at the tips of its sometimes 5.0-meter-high branches.

Saguaro cactus is attractive to animals, particularly birds. The cellulose interior of the plant provides natural insulation, which maintains a steady temperature through the hot desert days and cool nights. The small white-winged dove is chief pollinator of the giant saguaro blossom in the springtime and distributor of its seeds later in the year. There is great competition for accommodation in the saguaro among screech and elf owls, purple martins, and crested woodpeckers. The starling, mistakenly introduced from Europe to eastern North America, has now reached the Sonoran Desert and is threatening the very existence of some of the less aggressive indigenous birds.

result is a rhythmic geyser eruption on the surface of the basin.

If water from such a basin circulates beneath its surface without restriction, then, after being heated to a high temperature at its lowest point of circulation, the water will naturally rise up toward the surface of the basin and will emerge as a hot spring. Superheated water dissolves minerals from the volcanic rocks through which it circulates, and such mineral-water hot springs frequently form large near-boiling pools. Pools of this character are probably the nearest thing on Earth to Charles Darwin's concept of ". . . some warm little pond, with all sorts of ammonia and phosphoric salts, light, heat, etc. present, that a protein compound was chemically formed ready to undergo still more complex changes. . . ."

The most variegated pond to have nurtured life's beginnings on Earth over 3,000 million years ago could not have been more colorful or more exemplary of life's development than is the Grand Prismatic Spring above the Firehole River on the Yellowstone caldera. Grand Prismatic is a large spring, almost perfectly round (100 meters across), and its colors start with indigo-blue near-boiling water in the center that shades to cooler aquamarine toward the water's edge. As the heavily mineralized water surface gives way to a thin, surrounding, crustal deposit of golden-brown travertine, a veritable spectrum of life encircles the whole, a progression of emerald-green, yellow-ocher, and brilliant-red life forms: blue-green algae and filamentous bacteria. Above them all hangs a prism, a permanent cloud of evaporated water that splits the light reflected from the surface of this "warm little pond" into a spectrum of pale blues and greens and yellows.

All forms of life have to be either thermophiles, mesophiles, or cryophiles. Most of those blue-green algae and filamentous bacteria in Prismatic Spring are thermophilic — they thrive in high temperatures. We human beings and the other mammals are mesophilic — our lives depend on the middle range of life-permitting temperatures. Antarctica has a great variety of cryophiles, organisms that thrive in subzero conditions. But why such a wide range of temperatures in which life exists, and why are there so many types of organism in each of the three categories? The best available explanation is that every form of life is a product of the environment that existed at the time of its first appearance and that life in the general sense is far more determinedly rugged and adaptable than most of us imagine it to be. On the grand scale life is not tenuous, nor fragile, nor tentative; it is striving, responsive, and perhaps indestructible. Life's evolution is its track record of successful adaptation both to gradual change and to catastrophic change. More often than not, adaptation implies adaptation to changes in environmental temperature, with all that that statement implies: Suddenly an environment is too hot or too cold, and so there is too little water or too much water or too much ice for the continued existence of a species. If that species cannot move to a region where the environment is similar to the one to which it has grown accustomed, the species has to adapt to the new conditions or perish.

The fish populations in parts of the Pleistocene lake system in the Basin and Range Province had to meet just such a challenge to their continued existence. Some lakes in which the fish lived were suddenly (in evolutionary terms) reduced in volume by evaporation. These lakes became saline and hot instead of remaining fresh and cool. In many cases landlocked lakes were reduced to evaporite pans with an occasional briny pool tucked away in some shady corner. Streams still ran into the lake basins; and before the water evaporated (or trickled into the Colorado River system), they dissolved some of the salts that had accumulated on the desert surface. The streams became hot and uninhabitable — or so one would expect. Yet in Death Valley today there are several distinct and unique species of small fish, called "pupfish" because of their frisky behavior, which originated from a common stock of freshwater fish that populated lakes in the Death Valley region thousands of years ago.

Tarantula: Female tarantulas may live as long as 25 years. The males don't survive so long because females tend to devour their mates. Contrary to fable the tarantula is a fragile, rather gentle, animal; it cannot jump, it seldom bites human beings, and if it does, the bite hurts but is harmless. The tarantula can discard short barbed hairs if provoked, and these cause irritation if they penetrate the skin. The spider's principal enemy is the female tarantula hawk wasp, which stings the spider, drags its paralyzed victim to its hole, and lays its eggs on the spider's abdomen. When the eggs pupate, the larvae feed upon the still-living spider.

Western diamondback rattlesnake: This is a common snake in the Basin and Range. It grows to a maximum length of 2.0 meters. These snakes are just as effective striking their prey at night as they are during the day because they have heat-sensitive infra-red detectors, pit organs, below and in front of their eyes. They bear their young live, up to twenty-four at a time. One of the rattlesnake's enemies is the king snake, which is generally immune to rattlesnake venom. It constricts its victim and devours it head first. But the rattlesnake's main enemies are birds of prey, particularly the comic-looking roadrunner.

Ocelot: One of the most beautiful animals of the desert, the ocelot is about twice the size of the average domestic cat and has much in common with it. It hunts the brush-covered regions of the desert at night and is an extremely good climber. It preys on birds, small mammals, and snakes.

GRAND CANYON OF THE YELLOWSTONE RIVER, WYOMING

Extremely rare speciated descendants of this stock now live in very saline shallow streams and pools. At two separate locations in Death Valley pupfish live in stream water that is 4.6 times more salty than seawater and that varies in temperature from near freezing in winter to 40°C or more in summer. Other species of pupfish at Devil's Hole in the Amargosa Desert, 65 kilometers southeast of Death Valley, are believed to have become isolated very much sooner than their cousins in Death Valley did. These fish have adapted to an annual steady-state temperature of 33°C in the depths of a deep pool in the mouth of a limestone cave system that receives direct sunlight for only limited periods of the year. The pupfish are yet another new species of fish adapted to radical changes in temperature and salinity in a period of a few thousands of years, not the millions of years that Darwin had thought necessary to cause such speciation.

Pupfish are the progenitors of the future populations of fish that will evolve when next the lakes refill: Provided that the pupfish survive man's present intrusion, future freshwater lakes in the Basin and Range will be populated by other and different species of fish, which will have readapted from saline, high-temperature conditions to fresh, cool water. In this event the fossil record of a million years hence will puzzle some future paleontologist, who will read the hiatus but who, unless he is fortunate enough to find the extremely rare Death Valley species of pupfish, will miss the link between the two evolutionary events.

Other remarkable adaptations from moist to extremely arid conditions are provided by the evolution of the kangaroo rat, the camel (a pre-Pleistocene North American invention), and other mammals that live without the need to drink water or, if they *do* drink, have rare

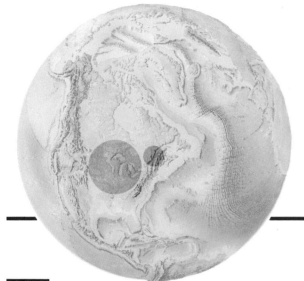

THE twin magma chambers beneath Yellowstone National Park last blew their tops about 600,000 years ago. The thin continental crust above them swelled. As a consequence a series of fractures formed around the two highest points of uplift. When the magma could no longer be contained by the weight of the weakened crust, there was an explosion of unbelievable magnitude. In hours many hundreds of cubic kilometers of rock fragments (breccia lava) and pumice were plastered over thousands of square kilometers of the surrounding landscape. What was left of the surface above the mantle plume collapsed, perhaps 1,000 meters into the void created by the explosion, plugging the two magma chambers. Yellowish rhyolite lava seeped through the concentric rings of faults and flooded the two overlapping calderas. To-

gether they now form the Yellowstone caldera, a roughly oval basin of some 3,500 square kilometers. All this happened in the Pleistocene Epoch about three-quarters of the way through an ice age. The still simmering cauldron was eventually filled with ice 1,000 meters thick.

The Yellowstone caldera is astride the North American continental divide, about 2,500 meters above sea level. This may not be just coincidental. Present, past, and future Yellowstones (like a string of Hawaiian Islands) form above a mantle plume with a huge body of hot and therefore swollen continental crust above them. As the crust moves southwest beyond the hot spot, the crust cools and shrinks and therefore reduces in height while the part above the mantle plume remains swollen and elevated. But hundreds of thousands of years ago the elevated Yellowstone caldera became the focal point of several con-

verging glacier systems from neighboring mountains. During this stage of Yellowstone's history thermal activity continued unabated beneath the ice, an activity similar to the present subglacial volcanic activity in Iceland.

Yellowstone Lake, which formed in the caldera from melting glaciers, was balanced on a sort of fulcrum: It might have discharged into either the Pacific or the Atlantic oceans. It did both. To the south it is still tenuously connected by a stream to Heart Lake, the headwater of the Snake River, which flows southward in front of the Grand Tetons, around their southern perimeter, and then down to the Snake River Plain eventually to join the Columbia River, which flows into the Pacific. But Yellowstone Lake's main outflow is to the north, where the river that cut the Grand Canyon of the Yellowstone joins the Missouri

Breida Lake, formed from the melting ice of Vatnajökull in Iceland. The ice front and the ice field behind it are not part of a glacier in the mountain-glacier sense; for Vatnajökull is a receding ice cap, a remnant from the last Ice Age. It is situated over an active volcanic region, part of the Reykjanes Ridge. Some typically flat-topped and subglacially formed volcanoes can be seen on the horizon. From events here it is possible to surmise the kind of catastrophic event that may have occurred in the Yellowstone geyser basin while it was at times under 1,000 meters of ice during the Pleistocene Epoch.

Thermal activity under Vatnajökull causes ice bursts (*jökullhaups*). These happen every few years in the region of Breida Lake. When the volume of trapped subglacial meltwater is sufficient, part of the ice cap is raised from the bedrock, and the accumulated meltwater escapes in a wall of water up to 10 meters high. This wave carries rock and debris to the outwash plain to the left of the picture and for many kilometers beyond. This is but one of the numerous ways in which ice and melting ice have shaped landscapes.

BREIDA LAKE, ICELAND

The Icelandic word jökullhaup is now broadly used to describe any glacial outburst — not just those caused by thermal activity. In Scotland, for example, an ice-dammed lake in Glen Roy and Glen Spean burst several times between 10,000 and 11,000 years ago, when the lake rose above a critical level; the glacial-lake levels can be clearly seen etched into the mountainsides of the region today. According to glaciologists at Edinburgh University the catastrophic flooding that resulted from these jökullhaups caused fluvial deposits to block what is now Loch Ness at both ends: at Beauly Firth (Inverness) to the northeast and at Fort Augustus to the southwest. The Rivers Ness and Lochy were cut by water overflow after these deposits were formed. The Loch Ness monster must have had a most uncomfortable time during these events!

The most devastating jökullhaup known was that which discharged

glacial Lake Missoula, 80,000 square kilometers in extent and about 600 meters deep (2,100 cubic kilometers of water), in a matter of days between 15,000 and 18,000 years before the present. This lake had formed behind a huge ice dam that had blocked the Clark Fork River between Montana and Idaho north of Yel-lowstone. The flood (40 cubic kilometers per hour) that resulted from the jökullhaup, during which the ice dam was completely destroyed, is estimated by the USGS to have been the equivalent of ten times the flow of all the rivers in the world combined. The Missoula discharged 11 million cubic meters per second; the Amazon discharges 170,000 cubic meters per second.

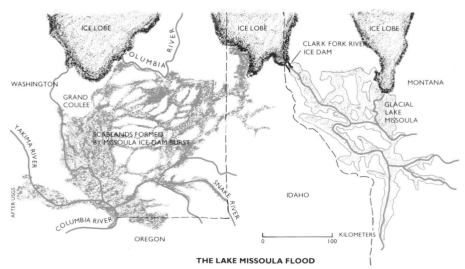

THE LAKE MISSOULA FLOOD

BERING LAND BRIDGE

CORDILLERAN ICE SHEET

CENTER

CENTER

KEEWATIN CENTER

PRESENT HUDSON BAY

LABRADOR CENTER

GLACIAL LAKE AGASSIZ

MAXIMUM GLACIAL ADVANCE

PACIFIC OCEAN

ANCESTRAL MISSISSIPPI RIVER SYSTEM

ATLANTIC OCEAN

GULF OF MEXICO

CARIBBEAN SEA

THE ILLINOIAN ADVANCE

⟵ Direction of ice movement

⟶ Minimum sea level

KILOMETERS
0   500   1,000

PANAMANIAN LAND BRIDGE

AFTER ROBERT H. DOTT ET AL.

Although comparatively rare in Earth history, there have been several ages of ice. The last one, the Pleistocene, started about 2.0 million years ago and ended about 10,000 years before the present. It is perhaps more correct to say that 10,000 years ago the Northern Hemisphere went into a glacial recession which seems to be continuing but which may reverse.

Greenland's huge ice cap is the largest Pleistocene remnant associated with the North American continent. But the Greenland ice cap was only a segment of a vast ice sheet sometimes 3,000 meters thick. This ice sheet once covered the whole of the nucleus and about two-thirds of the continental stable platform. This map shows the ice sheet at its maximum, a stage called the "Illinoian advance." European glacial periods occurred roughly simultaneously but have been given different names (see page 144).

It was during the Illinoian advance that the lobes of the advancing ice sheet established the main river-drainage patterns of the interior stable platform. The map accents the ancestral western rivers (the Missouri and the Arkansas), the ancestral eastern river (the Ohio), and the central-platform river (the Mississippi).

River, the principal tributary of the Mississippi, which discharges into the Gulf of Mexico.

The time had come to part from Martin Litton and his Cessna 195 — but not before an exhilarating flight over the Yellowstone caldera and among the strikingly sharp peaks of the Tetons. I had seen both Yellowstone and the Tetons many times, but never on so clear and scintillating a day. And to end the Litton segment of my journey on the same "one-of-those-days" note on which it had begun in San Luis Obispo, I had left my camera in the plane's storage locker.

On the northwestern flanks of the Yellowstone caldera are the headwaters of the Madison and Gallatin rivers, which at Three Forks in the Northern Rockies in Montana join the Jefferson River to become the great Missouri River, the "Big Muddy," so nicknamed because of its heavy load of suspended sediments. The Missouri first flows directly east from its Rocky Mountains outlet, parallel to the Canadian border. Just south of the former exit from the Alaskan Refuge the river's course changes to southeast and holds this general direction for 3,000 kilometers until its confluence with the Mississippi River north of the city of St. Louis. Here it is a little more than 100 meters above sea level and still 2,000 winding kilometers from the Gulf of Mexico.

Big Muddy once flowed northeast to Hudson Bay, but lobes of Pleistocene ice blocked its path. During the onset of the Ice Age the Laurentide Ice Sheet, as the North American Pleistocene ice sheet is called, thickened and developed amoeba-like lobes of ice that crept insidiously southward until lobes and body (sometimes 3,000 meters thick) enveloped two-thirds of the stable platform. For every 300 meters of ice thickness the continental crust was depressed 100 meters into the asthenosphere. The sallies of the Laurentide glacial advance and the whims of its retreat determined the drainage system of the craton, the nature and distribution of its soil, the undulations of its landscape, and thus the very destiny of all forms of life living upon it.

It seemed incredible to me that a little more than a century ago, almost yesterday, the proposal that continental ice sheets had once existed had been a matter of ridicule. This was strictly a European quarrel unintentionally provoked by Louis Agassiz, the young president of the Swiss Society of Natural Sciences. In 1837 Agassiz brought to the attention of the scientific elite of his day the long-known fact that there were some odd-looking boulders in the Jura mountains around Neuchâtel. The provocation lay in Agassiz's suggestion that these rocks and the debris associated with them may have been moved from their place of origin by glacial ice long since melted. Agassiz, himself a recent convert to the theories of his countrymen Jean de Charpentier and Ignace Venetz, asserted that there was much other evidence in the glaciated mountains and valleys of Switzerland that supported the idea of an ice age in the recent past. There followed a dispute that lasted more than twenty-five acrimonious years. The general belief of that day was that such "erratics" — boulders that hadn't originated where they then lay — were evidence of the Noarchian Flood and that to suggest differently, much more to argue the point, was blasphemy. Yesterday's most ardent converts are often today's most powerful advocates, and so it was in Agassiz's case. The real assault on established ideas began in l840, after Agassiz addressed a Glasgow meeting of the British Association for the Advancement of Science, a meeting that was attended by Charles Lyell, the Reverend Professor William Buckland, and other lions of the great Victorian age of science. In fact, after a field trip in Scotland with Agassiz and Murchison (of the Cambrian-Silurian dispute that I previously mentioned), it was Buckland who next became both a convert to, and a most powerful British advocate of, Agassiz's ideas.

In 1852, more than half a century after Captain Cook's neglected Antarctic discoveries, James Ross, in command of H.M.S. *Terror* and *Erebus*, explored the east coast of Victoria Land and thus established that Antarctica was an ice-covered continent. In the same year it was understood for the first time that Greenland also has an enormous ice cap. Meanwhile Agassiz had carried his assault on geological convention to North America and in particular to the ice-scored rocks of New England. Agassiz's enthusiasm for his subject captivated John Amory Lowell, a wealthy Massachusetts

textile-mill owner, who ensured that his protégé received a professorial appointment at Harvard University. Massachusetts became Agassiz's permanent home. Before his death in 1873 Agassiz's theories had been broadly accepted by the geological profession, and by mapping terminal moraines of long-departed glaciers, geologists worldwide had begun assembling information that, only two years after Agassiz's death, made possible the publication of a global map of Pleistocene glacial advances. The map showed that the Pleistocene ice sheets were mainly a northern hemisphere phenomenon (although there had been a global change in climate) and that Agassiz had been wrong in one important respect:·The ice cover had not been a simple extension of the North Polar region, as he had thought, but had consisted of a series of individual ice caps sometimes joined together.

There proved to be two classic cases of large-scale depressions, or basins, which had been formed by ice sheets and which had rebounded isostatically when the ice melted. Both were coincident with continental shields. The first is the Fennoscandian Ice Sheet, which covered Sweden and the Gulf of Bothnia and which caused the similarity between the resulting glaciated topography of Sweden and Minnesota — and perhaps also explains the attraction of that state for Swedish immigrants. The second and much more extensive ice sheet was the Laurentide, which determined the course of the Ohio River as well as the Missouri, re-established the headwaters of the Mississippi, formed the Great Lakes and other even greater lakes now gone, shaped the Gulf of St. Lawrence, and provided the prime ingredient for the soil that makes the Canadian Prairies, the Great Plains, and the Mississippi basin the most productive agricultural regions on Earth.

When the Fennoscandian and the Laurentide ice sheets melted, the removal of their weight caused the continental nuclei on which they had accumulated, the Canadian Shield and the Baltic Shield (which is centered beneath the Gulf of Bothnia), to bob back to the

The Gulf of Bothnia and the Baltic Sea together parallel Hudson Bay. Both regions are basins formed on continental shields: the Baltic Shield and the Canadian Shield. Both regions were the focal points of Pleistocene ice sheets; and as shown here, both areas rebounded and are still rebounding after the heavy ice had melted. Both basins are tending to tilt more toward the south as they uplift. In both cases the continental cratons were depressed to a depth of about one-third of the thickness of the overlying ice.

FENNOSCANDIAN ICE-SHEET REBOUND
Former shorelines around the retreating Scandinavian Ice Sheet

AFTER. ATLAS OVER SVERIGE

LAURENTIDE ICE-SHEET REBOUND
Contours indicate rate of uplift in meters per 100 years

AFTER ROBERT H. DOTT ET AL.

PLEISTOCENE TERMINOLOGY

IN NORTH AMERICA
(Upper Mississippi Valley)

IN EUROPE
(Alps and Southern Baltic Region)

IN NORTH AMERICA	IN EUROPE
WISCONSONIAN GLACIAL	WÜRM GLACIAL
Greatlakean Advance (8,000–11,000)	Younger Dryas Advance (10,000–11,000)
Twocreekian Retreat (11,800–12,500)	Allerod Retreat (11,000–12,000)
Woodfordian Advance (12,500–22,000)	Older Dryas Advance (13,000–20,000)
Farmadalian Retreat (22,000–28,000)	Riss-Würm Retreat (20,000–30,000 +)
Altonian Advance (28,000–70,000 +)	Riss Advance (30,000–60,000 +)
SANGAMONIAN INTERGLACIAL	MINDEL-RISS INTERGLACIAL
ILLINOIAN GLACIAL	MINDEL GLACIAL
YARMOUTHIAN INTERGLACIAL	GUNZ-MINDEL INTERGLACIAL
KANSAN GLACIAL	GUNZ GLACIAL
AFTONIAN INTERGLACIAL (600,000–700,000)	DONAU-GUNZ INTERGLACIAL
NEBRASKAN GLACIAL	DONAU GLACIAL
VOLCANIC ASH (2,200,000)	
PRE-NEBRASKAN GLACIAL	

Numbers indicate years before present, where known.

surface of the Earth rather like surfacing submarines' discarding water as they rise from the depths and — also like surfacing submarines — oscillating before they stabilize. The rate of uplift within the shield boundaries and the overlying stable platform varied, as did the locality of uplift, the angle of the tilt caused by uplift, and the rate of ice accumulation and melting — all these factors combine to create an extremely complex picture, which is made all the more complex by a maze of names coined by geologists since Agassiz's time to describe the different lakes that formed on both continents and the different stages of glacial advance and retreat in Europe and North America. The last of the North American glacial advances is the Wisconsin, and indeed at the University of Wisconsin in Madison are many of the leading scientists who specialize in the Pleistocene Ice Age. Robert H. Dott, Jr., James C. Knox, and John E. Kutzbach separately and most patiently talked with me about their respective fields of research.

Today the rebound of both the Canadian and the Baltic shields is approximately 1 meter or less per 100 years. Both shields will eventually stabilize and may have already done so in certain localities. Lakes that formed in melting ice-lobe depressions, such as the Great Lakes and the Gulf of Bothnia basins, radiate in a general fashion from a center point on the shields on which the ice caps formed. As the shields rebound, so the lake basins gradually tilt away from the center point of rebound. In the case of both the Great Lakes and the Gulf of Bothnia this tilt is generally southward. The present waterline levels on northern shorelines in both regions are therefore lower, and the southern-shoreline water levels higher, than they were in the past; Lakes Superior and Ontario, for example, are increasing their tilt to the south at about 20 centimeters per century. Although smaller in overall extent, the Gulf of Bothnia is the European equivalent of the Great Lakes: It was formed in the same way by the same processes at roughly the same time. The difference is that the Gulf of Bothnia has a direct sea link via the Baltic Sea to the North Sea, while the Great Lakes don't have a sea-level outlet.

At the end of the Wisconsin advance, which concluded between 8,000 and 11,000 years ago, the melting ice lobes in the Great Lakes basins and in the Winnipeg and Manitoba lake basins to the northwest of them first formed small prototype lakes at the foot of each main lobe. (These are called "postglacial lakes," and since geologists give each lake a different name for a different stage of its development, to prevent confusion, the present familiar names are used as far as possible.) As glacial retreat continued, the lakes grew to a considerable size. Many of them joined to form even bigger lakes, which drained mainly into the Mississippi basin. Meltwater from the eastern lobes of the ice sheet, which formed part of the Erie and Ontario basins, drained eastward down the present Mohawk River and then abruptly south down the fault line of the present Hudson River from Albany and so to the Atlantic Ocean. At this time the present St. Lawrence River drainage was blocked by the Laurentide Ice Sheet, which overlapped the entire length of the present river and gulf.

In the next main stage the Laurentide Ice Sheet continued its generally northward retreat, leaving two vast lakes in regions to the south of Hudson Bay, which previously had been covered with up to 3,000 meters of ice. One of these postglacial lakes, which was larger in extent than all the present Great Lakes combined, was named after Agassiz. Lake Agassiz drained into the Mississippi basin, leaving the present Lakes Winnipeg and Manitoba as residual lakes. The other main postglacial lake was a bloated version of the present Great Lakes, and it too continued to drain mainly into the Mississippi basin and to a lesser extent down the Mohawk and Hudson valleys.

The final stage of Great Lakes evolution is the most controversial, for it was complicated by rebound, changing lake levels and flow directions, glacial resurgence, and possible sea incursion (the Champlain Sea). About 12,300 years ago the Laurentide Ice Sheet's retreat uncovered the fault followed by the St. Lawrence River, thus leaving a Lake Ontario much larger than today's free to drain eastward beyond Montreal; but because ice still filled the basin immediately below the present city of Quebec, the high level of water in Lake Ontario forced the lake to drain southward from what is now the Trois-Rivières region east of Montreal. The outlet followed the line of the present Richelieu River, the Champlain Valley, and the Hudson River to the Atlantic. When the foot of the retreating Laurentide Ice Sheet in the Quebec basin and east to the Gulf of St. Lawrence had melted sufficiently, Lake Ontario's water could escape into the Gulf of St. Lawrence. This *might* have been a catastrophic event like the flood of Lake Missoula in the western region of the continent. Whatever the circumstances, the level of Lake Ontario fell sharply, and the Hudson River-Lake Champlain route was abandoned, leaving that lake as a residual freshwater lake.

The St. Lawrence had gained the upper hand; it was now an established river. Until now (about 11,000 years ago) there had been a broad neck of land separating Lake Erie from an enormous Lake Ontario. The two lakes were joined by the Niagara River. As the level of Lake Ontario fell, the Niagara became a cataract that began to cut back into the Niagara escarpment. The cataract became a floodgate, and the present unrelenting progress of Niagara Falls toward the capture of Lake Erie began.

This new drainage, preferred because of its short, steep, and therefore fast erosion rate, gave the lakes direct access to sea level in the Quebec basin. This, together with the tilting effect of the rebounding craton, started a domino effect that established the present general level of Lake Erie and caused one Great Lake to overflow into the next in a generally eastward direction, thus reducing the flow to the Mississippi drainage basin and gradually allowing the lakes to assume their present familiar maple-leaf outline. Eventually no connection remained between the Great Lakes and the Mississippi basin or the Hudson River. The very short and geologically very young river, the St. Lawrence, became the main outlet for the whole Great Lakes system.

This was quite a geological coincidence, because the fault followed by the St. Lawrence River formed as the result of the collision of Laurentia with Baltica (ancestral North America and northern Europe) during the early stages of the assembly of Pangea, about 440 million years ago.

It was bitterly cold in the Cité de Québec, a special brand of imported European cold, with an icy windblown rain added to guarantee chilling of the marrow. Icebreakers were moored at a quay beside Boulevard Champlain. Sardonically, I wondered whether they would be needed. There was to be no photography that day; but when it *was* completed, Joy and I intended to drive along the shores of Ontario, Erie, and Michigan to Chicago and then to follow the course of the Illinois and Mississippi to St. Louis, Missouri, and to New Orleans on the Gulf of Mexico. There was nothing new about the intention: The journey had been done centuries before by more stoic adventurers in canoes on lakes and rivers. My purpose was to follow their lead, to see for myself. The time seemed opportune to pore over maps in a warm hotel room, make a few phone calls, and digest what I had learned about ice sheets and climates during a few days at Madison, Wisconsin, on the way to Quebec. And then perhaps to sample the *haute cuisine* of the Continental Restaurant in old Quebec (which was superb). Loew's La Concorde had provided a room with a superlative view, a reminder of the Captain Cook Hotel in Anchorage. This time instead of Cook Inlet and the Aleutian Range the picture

This sketch shows the approximate size and geography of the Laurentide Ice Sheet at a late stage of its retreat between 10,000 and 15,000 years before the present. The northeastern Laurentide retreat was slower than the southwestern retreat was: The flow of water was therefore mostly toward the Mississippi basin. Glacial Lake Agassiz (residual Lake Winnipeg) was the largest lake by far ever to have formed on the stable platform during the Pleistocene Epoch. The northeastern drainage was first via the present Mohawk and Hudson rivers and later, after the retreat of the ice sheet, via the present Lake Champlain to the Hudson River and the Atlantic.

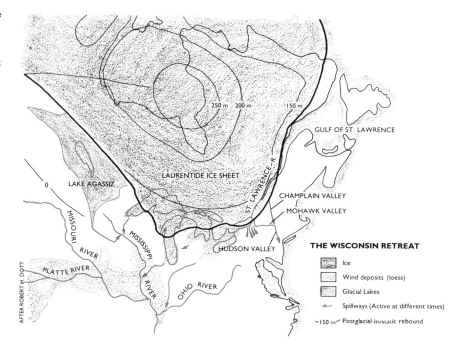

THE WISCONSIN RETREAT

Ice
Wind deposits (loess)
Glacial Lakes
Spillways (Active at different times)
~150 m Postglacial isostatic rebound

NIAGARA FALLS, NEW YORK

When the Gulf of St. Lawrence and the St. Lawrence River at the present Quebec were at last free from Laurentide ice (see diagram on page 145) and when the craton had rebounded sufficiently from the weight of this ice, the Great Lakes region stopped discharging into the Mississippi basin: the Mohawk River and Lake Champlain routes of drainage were abandoned so that there no longer was a connection between the Great Lakes and the Hudson River. When Lake Ontario had dropped to approximately its present level, the Niagara River began cutting its way back into the escarpment that separated Lakes Ontario and Erie. That was 11,000 years ago, and the result of the cutback is Niagara Falls.

The immediate impression one gets at the very edge of Horseshoe Falls (seen here) is of overwhelming power. There are other waterfalls that are higher or wider or noisier. At the Canadian Horseshoe Falls one is simply hypnotized by the fast-curving green mass of perpetual, inevitable, and unstoppable water pounding itself to atoms.

Then one is surprised — surprised to discover that the American Falls are kept alive artificially by redirecting the river toward them from upstream. If left to themselves, they would by now be a trickle, abandoned by the river.

window overlooked the Quebec basin, Château Frontenac, the Plains of Abraham, the Laurentian Mountains, and the upper reaches of the mighty St. Lawrence. It seemed the right setting in which to think through what I had learned from Bob Dott, Jim Knox, and John Kutzbach.

In a sense ice sheets are subject to business pressures. Just as long as they stay within their precipitation "budgets" (the professional geological term), ice sheets stabilize or grow.

Their "income" (my terminology) in way of snowfall must exceed or equal their expenditure of snowmelt and evaporation, or they will suffer a deficit and go into decline. If a shortfall is not remedied in time, if the snow business climate continues to be unfavorable, then within the constraints of their working capital ice sheets inevitably become insolvent and go out of business. It follows that, although a large ice sheet (like a large business company) is to some extent exponential, creating its own weather and therefore its own momentum, neither ice sheet nor business company is self-perpetuating.

As an ice sheet grows in thickness by accumulating snow on its upper surface, which turns to ice under pressure from further snowfall upon it, the sheet spreads from its high points outward; for ice is naturally plastic. The direction of an ice sheet's flow is determined by the angle of inclination of these high points. An ice sheet tends to flow in all directions at once; but it flows fastest where it is steepest and little or not at all where its surface is level or depressed. Unlike a mountain glacier, the flow of an ice sheet may be influenced

QUEBEC, CANADA

but is not controlled by the inclination of the rock surfaces on which it moves. The Laurentide Ice Sheet was up to 3,000 meters thick — an almost inconceivable thickness and weight — but it was still the crests of the amorphous sheet and the complex motions of its internal layers that determined its amoeba-like movement.

The great pressure of an ice sheet on the rock surface beneath produces a lubricating interface of water similar to the interface between an ice-skate blade and the ice. As the sheet moves over the bedrock, it scrapes the surface, incorporating rock debris. The debris-loaded ice grinds the bedrock surface into a large variety of familiar glacial landforms. These landforms have prompted a special language to describe them, a nomenclature even stranger than the language of the tectonic scientists.

A mix of powdered rock, gravel, and hand-sized pieces of rock is termed "glacial till." Enormous quantities of till are dispersed within the mass of an ice sheet by the mixing effect of internal laminar flow. Compacted till beneath the sheet forms elongated hills called "drumlins," and small depressions beneath the sheet become filled with stratified sand and gravel, forming "kames." Part of the till is deposited in long sinuous ridges, "eskers," created by fast-flowing rivers tunneling in stagnant ice toward the periphery of the sheet. An ice sheet tends to flow around an obstruction, carving off pieces as it does so to form "monadnocks" and smaller obstructions, hillocks of extremely resistant rock called "roches moutonnées" (sheeplike rocks!).

The progressive lobe of an ice sheet continues to nose its way forward until the resistance it meets is greater than the force it exerts, in which case the lobe stops moving forward or

changes direction, leaving a deposit of till to mark the limit of its traverse. Eventually conditions of climate are reached by the southern progression of a lobe that cause the rate of its melting to exceed its rate of advance. The ice front of the lobe has thus reached a state of equilibrium — but the sheet behind it continues to advance. The result is the accumulation of a frontal moraine (which may or may not be terminal), on which the ice sheet dumps its load of till. The meltwater from the ice front forms braided streams, which create outwash plains. These drain into a river, where the silts form "braid bars" in the upper reaches.

When an ice sheet recedes, large blocks of rotting ice become detached from it. These may sink into soft surfaces and deposit till as

MONTMORENCY FALLS, QUEBEC

The fault line followed by the St. Lawrence River (here upstream toward the city of Quebec on the left-hand horizon) is the geological boundary between the Grenville Province and the Northern Appalachians, between the geographical provinces of Quebec and New England. The 84-meter cliff over which the Montmorency River flows is literally the exposed edge of the Canadian Shield, which forms the north shore of the St. Lawrence. The river flows in a fault that resulted from the collision of the pre-Pangean continents of Laurentia and Baltica about 440 million years ago.

The promontory on which the delightful city of Quebec stands today, Cape Diamond, was discovered for France in 1535 by Jacques Cartier, the sailor-explorer commissioned by the French king, Francis I, to discover a route to the Orient. When Cartier arrived, Huron Indians occupied the village of Stadacona beneath the cliff. Cartier had planned to "experience a Canadian winter," and this he must have regretted. In addition to suffering from extremes of cold, the likes of which he and his crew had not previously experienced, he lost many men from scurvy and other diseases. After weeks of distress, Cartier's men were "miraculously" cured in days by a Huron brew prepared from the white cedar tree that thrives there on the limestone outcrops and poorly drained ground.

The name of Quebec is derived from an Algonquin Indian word meaning "narrowing of the river." It is here that the freshwater St. Lawrence meets the tidewater passage that leads to the Gulf of St. Lawrence and the North Atlantic. Because of Point Diamond's commanding position at this gateway to the North American continent, Quebec was for many years the battle prize of, and besieged by, both British and French forces. Eventually the city surrendered in 1759, after a bloody conflict in which General James Wolfe's British forces successfully outflanked the Citadelle at night by rowing upriver and attacking from the rear (beyond the headland in the picture). Both Wolfe and the opposing commander, the Marquis Louis de Montcalm, were killed in the battle.

Largely dependent on the fur trade, the French Province of Canada became the British Colony of Quebec, which was expanded in 1774 to include all the inland territory of Canada to the Ohio and Mississippi rivers. French and English were both official languages, together with French civil and British criminal law. Quebec became a cultural and administrative center, and the Château Frontenac (now a hotel) a famous meeting place: Franklin D. Roosevelt and Winston S. Churchill, for example, met there twice during World War II to plan the invasion of Normandy, a Churchillian touch.

The picture shows the old city (*vieux quartier*) of Quebec, the Citadelle (above the cliff to the left), the Château Frontenac Hotel (center), Séminaire de Québec (1663, the original Laval University site) at extreme right, icebreakers by the quays along Boulevard Champlain, and Loew's Concorde Hotel tower in the center-horizon background.

they melt, creating an impervious concave surface that can form isolated lakes with no apparent water supply. These lakes are called "kettles," or "kettle lakes," and there are hundreds of thousands — perhaps millions — of them left in the wake of an ice sheet's retreat. They can be from 10 meters to 10 kilometers in diameter, and 20 meters or more deep. Such large kettles or well-kettled areas, or even large and less well-defined depressions that have ponded, slowly fill with sediment and become extensive marshlands. During periods of steady glacial recession the ice sheet also deposits an even covering of till on desolate landscape; but if the recession is reversed by a glacial resurgence, a new frontal moraine will form where and when the glacial recession resumes. The most remarkable succession of such frontal moraines, many hundreds of kilo-

WISCONSIN DRUMLINS

WISCONSIN KAMES

WISCONSIN RIVER, WISCONSIN

Today's landscapes of the Northern Hemisphere's stable platforms were largely molded and shaped by invading and retreating Pleistocene ice sheets. These aerial pictures of Wisconsin (northeast of Madison) show the direction of one advancing lobe of the Laurentide Ice Sheet during the Wisconsin advance about 12,000 years ago (the road in the main picture runs due north) and secondly illustrate how the vicissitudes of an ice sheet have controlled human use of the landscape. The elongated shapes are "drumlins" formed from compacted glacial "till" as the advancing ice sheet moved across the landscape. Till is a mixture of powdered rock, gravel, and fist-sized pieces of rock, the main product of glacial erosion.

The irregularly shaped hills, called "kames," above, left, were fluvial deposits formed either at the margin of the ice sheet as it retreated or by sand and gravel deposited in a depression by a subglacial stream.

The picture (below) of the Wisconsin River (associated with Louis Joliet's 1673 expedition) near Devil's Hole, a terminus of the Wisconsin advance, illustrates a "braid bar." Braid bars are sandbars of fine material that accumulate and divide slow-moving glacial rivers. Such rivers are formed from water running off outwash plains, which are in turn produced by braided streams from water that issues from the glacier snout.

meters long and often much more than 100 meters high in ridge after ridge, were left at the southern end of all the Great Lakes. They are effective dams that naturally prevent the Great Lakes from discharging to the south.

One large part of the stable platform in Wisconsin that is within an otherwise heavily glaciated region remained free from glaciation during the last major advance. The region is called the "driftless area" even though large quantities of glacially derived sand and gravel outwash exist in the region. But since there is no other evidence of glaciation, it is believed that the amoeba-like spread of the Wisconsin ice sheet was discouraged by the local topography and didn't invade: There doesn't seem to be any other explanation for the driftless area's lack of drumlins and kames. The distribution of drift in unglaciated areas raises the issue of how this part of Wisconsin and many other driftless areas on the stable platforms of North America and Northern Europe acquired such rich soil. The answer is that wind separated and blew dust from the rough glacial drift lying on outwash and floodplains and literally blanketed the stable platforms with the dust. This material is extremely fine, and its particulate size enables it to hold large amounts of moisture.

Imagine a fist-sized lump of white chalk dipped for a few seconds in a pot of ink: When removed and cut in two, the ink will be seen to have penetrated only part of the way into the lump. If a similar lump of chalk is chopped into a large number of small pieces that are simultaneously dipped in the ink, each small piece will absorb ink throughout its mass. Obviously the fist-sized lump cannot carry more than a fraction of the moisture that can be carried by the same volume of smaller pieces. Increased surface permits a material to carry more moisture; finely divided material does not have to be porous to carry more than its own weight in moisture. Thus particle size is the key to many natural (and industrial) processes, and one of the most important natural processes of all is the formation of soils. The extremely small particles of silt in glacial till not only give moisture capacity to a soil but also enable important nutrient minerals to be leached from the silt — something that larger particles or crude lumps of rock could not supply. Silt and other substances, like sand-sized grains of rock and fibrous material, together with water and air interact chemically to produce a fertile soil. But the main ingredient is silt. On its own, silt is just fallow dust; with water it becomes mud; too much water and it becomes a muddy stream. Windblown drift that was ground as fine as the finest flour by the ice sheets of the Northern Hemisphere is called "loess," a German word. It is from this glacial windblown, eolian, silt that rich loamy soils and billowy topography are produced. The morphology and the fertility of large tracts of the North American stable platform and the destiny of life upon it have very much depended on this product of the Laurentide Ice Sheet (see page 145).

The causes of ice ages and the causes of global climatic change have long confounded scientists. The mystery appears to be partly solved as a consequence of the new intensity of oceanographic research stimulated by the concept of plate tectonics. This had followed the discovery of paleomagnetism, magnetic reversals, and ocean-floor spreading. After half a century of controversy these discoveries had vindicated Alfred Wegener's outrageous hypothesis of continental drift. Although there is not yet broad agreement about their conclusions, specialists working in the field of climatology are agreed that the once-rejected theories of the Yugoslav mathematician Milutin Milankovitch, a contemporary and friend of Wegener, have also been vindicated. The evidence in support of Milankovitch came from the same place as the evidence that supported Wegener: the ocean bed.

Examination of cores removed from the Indian Ocean in 1976 by scientists at the Lamont-Doherty Geological Observatory (James Hays, John Imbrie, and Nicholas Shackleton) established climate patterns for a period of 500,000 years into the past, almost back to the time of the Yellowstone caldera eruption described at the beginning of this chapter. Milankovitch had predicted that major glacial advances and retreats would follow variations in the Earth's tilt and distance from the Sun at the equinoxes. The Earth's behavior is mathematically predictable but nevertheless eccentric in its movement around the Sun. This eccentricity accounts for a twenty-second difference in the length of one year to the next and also, so Milankovitch claimed, for variations in the Earth's radiation income from the sun, which create a complex cycle accounting for ice ages and their fluctuations. Hays, Imbrie, and Shackleton found that major changes in global climate reflected in the content of the Indian Ocean sedimentary cores were in accord with patterns predicted by Milankovitch.

Yellowstone's hot-spot eruptions occurred 0.6, 1.2, and 1.8 million years ago. The enormous amount of dust that was distributed during each one of these eruptions was carried by prevailing winds to settle on parts of the Laurentide Ice Sheet or on bare glaciated or unglaciated terrain. The date and origin of volcanic dust can be verified accurately. Yellowstone dust of different ages and composition therefore acts as a marker for dating stages of the Laurentide Ice Sheet's advance and retreat. These and other date marks interpreted in conjunction with the Milankovitch Cycle have shown that, although there have indeed been four major advances and retreats, there have also been mini-advances and -retreats and mini-mini-ditto: a "big fleas have little fleas" situation. That is why it is so diffi-

cult to tell whether we are moving into or out of an ice age now.

But in Quebec, looking out at the wet, windy, and cold scene through the picture window of my room in Loew's La Concorde Hotel I felt little doubt about the present inclination of the climate. I could almost visualize the ice cap forming in the Quebec basin beyond Château Frontenac. I was once asked, "Where have all the glaciers gone?"; and before I could respond, I was told, "They've gone back for more rocks." In the winter of 1535 and 1536, if he had known about ice ages and glacial erratics, Jacques Cartier just might have shared my amusement, but I doubt it.

Cartier was a ship's captain, a Breton from St. Malo, who had been commissioned by Francis I of France to find a northern route to Asia. Francis had added the customary homily to Cartier's charter about the need to discover gold and spices en route. In the previous year, 1534, Cartier had ventured as far as St. Peter's Strait off the north coast of the island of Anticosti in the Gulf of St. Lawrence. He had determined that a channel of considerable width lay beyond and thought that this might prove to be the fabled Northwest Passage. He had taken captive two Canadians (Cartier's term for the aboriginal people) and now returned to the gulf, having cautiously appointed his captives, Huron tribesmen, to be the first St. Lawrence river pilots: They aided the passage of French ships fully manned and provisioned up the great river. After almost a month of sailing upstream, Cartier discovered that the broad, tidal, saltwater channel of the St. Lawrence narrowed to a freshwater river by the promontory on which Quebec now stands. Cartier must have felt at this point that his chances of reaching the Orient were beginning to look slim. In September, 1535, he gingerly explored upstream from Quebec to Mont Royale (now the site of Montreal) and to his dismay there found the Lachine Rapids. This extensive obstruction, perhaps the scene of some calamitous reduction of Lake Ontario when melting Laurentide ice had freed its flow eastward, made the river impassable to all

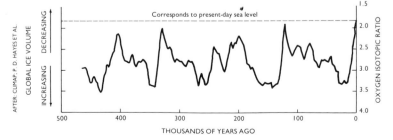

GLOBAL ICE VARIATIONS

transport, except canoes, until modest canals were completed for the use of specially built boats late in the 18th century.

By this time exploration of the Great Lakes drainage system had proved that it provided direct access to the very heart of the North American craton. It was by then understood that the Great Lakes were not just lakes but veritable inland seas and that, if advantage was to be taken of them, the problems of portage around rapids and between one lake and the next had to be solved by better means than birch-bark canoes, packhorse and wagon, or just shank's pony: The solution lay in the building of canals. In centuries past similar problems had been solved by the Phoenicians, Assyrians, Sumerians, Egyptians, and Romans. The construction of canals in far-off times had always been spurred by the need to solve the two most persistent problems of urban and military development: transportation and distribution where water supply was inadequate or rivers unnavigable or roads nonexistent. One of the oldest and best developed technologies was about to flourish in North America. The first canals to circumvent the Lachine Rapids in 1783 were shallow but effective. Their use was limited to flat-bottomed, tapering boats and canoes.

Suddenly the whole future scheme of things became apparent to the hard-headed American leaders of the time. The Industrial Revolution in Britain had stimulated the development of canal technology in the 18th century. George Washington had been a surveyor and had long recognized canals to be a possible solution to the young nation's great problems of distribution. And then in 1803 Thomas Jefferson, on behalf of the United States, had bought from France the half of the North American craton that lies west of the Mississippi River — the Louisiana Purchase of 1803. The potential of the Mississippi basin was considerable, but its potential value was incalculable if it could be linked to the Great Lakes by a canal system.

In the spring and summer of 1807 a series of intriguing anonymous letters was published in western New York newspapers. They were addressed to the public at large and were simply signed "Hercules." The writer suggested the building of a canal, the then technically improbable Erie Canal. The proposal was that New York harbor and the lower Hudson

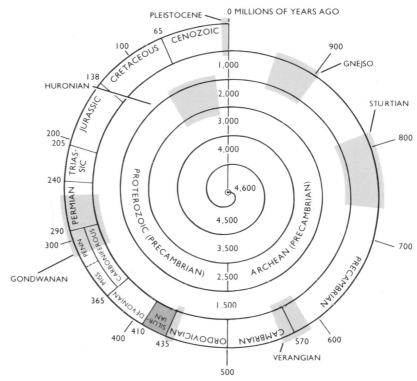

MAJOR ICE AGES THROUGH TIME

Some ice ages in the past were long in comparison with the 2.0 million years of the Pleistocene. Yet if the spiral in this diagram were unwound, the main glacial periods shown would be a small fraction of its total length.

The paleogeographic chart illustrates the approximate position of Laurentia and Baltica at about the time of the Taconian orogeny, the first of four mountain-building stages in the formation of the Appalachians. The St. Lawrence River follows a fault that resulted from this first stage.

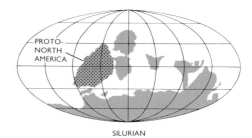

SILURIAN

BINGLEY FIVE-RISE LOCKS, YORKSHIRE, ENGLAND

LOCKPORT CANAL, NEW YORK

WELLAND CANAL, ONTARIO

The builders of the Erie Canal wanted a canal of a size and engineering difficulty well beyond anyone's previous experience. In 1817 they sent a young surveyor, Canvass White, to tour British canals (he is reported to have walked over 3,000 kilometers of canal bank in the process!). The British, already deep into the Industrial Revolution, had to build a network of canals to transport voluminous and heavy loads between industrial towns and ports in an age without adequate industrial roads and no railroads.

The American problem that caused most concern was how to surmount the Niagara escarpment, which separates Lakes Ontario and Erie. The British had already solved a similar problem in the Pennine hills between Yorkshire and Lancashire by building the Bingley Five-rise Locks (shown, as it is today, on the opposite page) on the Leeds and Liverpool Canal. The American engineers adopted the idea, but increased width and depth and built two tiers instead of one; the result was the Lockport, New York, double five-rise locks, which overcame the last hurdle before Buffalo and Lake Erie.

The Erie Canal was opened on October 26, 1825. It was not just successful: It was the vital link between the East Coast and the Central Plains. It led to an extraordinary development of agriculture, industry, and cities in the very heart of the continent.

The original Lockport double-tiered five-rise locks (above, left) have long been superseded. One tier of the original Erie Canal has been preserved, seen here on the right-hand side of the picture.

Above, right, is the first of eight locks in the Welland Canal, which links Lake Ontario to Lake Erie (45 kilometers) on the Canadian side of the Niagara River. The first Welland Canal (of three) was completed in 1829, and the present one was built in 1932 in anticipation of the construction of the St. Lawrence Seaway. In fact the Seaway, 3,500 kilometers long from the Lachine Rapids near

Montreal to Duluth, Minnesota, on the shore of Lake Superior, was not opened to ocean-going vessels until 1959. The reason for delaying the inevitable development for a century was simply that the eastern-seaboard cities that had prospered as a result of the Erie Canal and the American railroad companies with vested interests of their own were concerned for their future and therefore promoted tactics to delay completion.

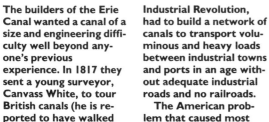

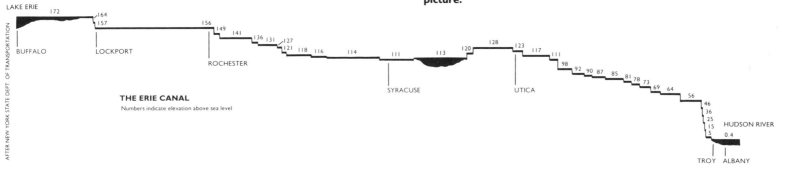

THE ERIE CANAL

Numbers indicate elevation above sea level

AFTER NEW YORK STATE DEPT. OF TRANSPORTATION

LAKE ERIE — 172 — 164 — 157 — BUFFALO — LOCKPORT — 156 — 149 — 141 — 136 — 131 — 127 — 121 — 118 — 116 — 114 — 111 — 113 — 120 — 128 — 123 — 117 — 111 — 98 — 92 — 90 — 87 — 85 — 81 — 78 — 73 — 69 — 64 — 56 — 46 — 36 — 25 — 15 — 5 — 0.4

ROCHESTER — SYRACUSE — UTICA — HUDSON RIVER — TROY — ALBANY

CHICAGO, ILLINOIS

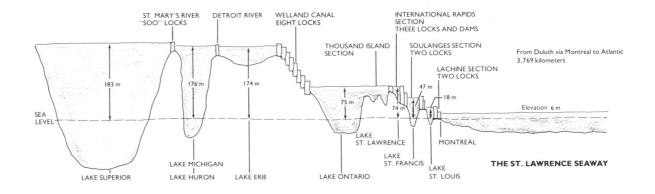

ST. MARY'S RIVER
"SOO" LOCKS

DETROIT RIVER

WELLAND CANAL
EIGHT LOCKS

THOUSAND ISLAND
SECTION

INTERNATIONAL RAPIDS
SECTION
THEEE LOCKS AND DAMS

SOULANGES SECTION
TWO LOCKS

LACHINE SECTION
TWO LOCKS

From Duluth via Montreal to Atlantic
3,769 kilometers

183 m

176 m

174 m

75 m

74 m

47 m

18 m

Elevation 6 m

SEA
LEVEL

LAKE SUPERIOR

LAKE MICHIGAN
LAKE HURON

LAKE ERIE

LAKE ONTARIO

LAKE
ST. LAWRENCE

LAKE
ST. FRANCIS

LAKE
ST. LOUIS

MONTREAL

THE ST. LAWRENCE SEAWAY

Louis Joliet had been commissioned by the governor of New France (Canada) to explore the Mississippi River in 1673. On his return upstream Joliet followed the northeasterly route of a Mississippi tributary, the Illinois River, to its head-waters on the west-sloping flanks of a now compacted and overgrown moraine. On the eastern slope of this shallow bank he discovered another river, the short, sluggish Chicago River, which flowed into Lake Michigan. Later Joliet suggested to his sponsor that a connection be-tween the two systems would provide an all-water link between Lake Michigan and the Mississippi. This was an idea that the new state of Illinois (1818) later endorsed, but although surveys were completed and several plans submitted, no funds were provided.

The completion and successful operation of the Erie Canal (1825) changed the status of the Chicago River from a diminutive swamp drain to a prospective transcontinental link between New York, the Mississippi, and the Gulf of Mexico via the west-flowing Illinois River. But some pioneering Chicagoans thought that a railroad system would do as well as, if not better than, a canal. After the customary vituperative arguments the outcome was the building of both the Illinois and Michigan Canal in 1848 and the Illinois Central Gulf Railroad.

As a consequence of a moraine formed during the Illinoian glacial re-treat from the center of a stable platform, one of the world's great cities was founded and was (in the mid-19th century) on its way to fame and fortune. This panorama shows the result: Chicago on the shores of Lake Michigan today.

should be connected, via Albany through the Mohawk River valley, to Buffalo on the Niagara River at the foot of Lake Erie, a route that had been well prepared by the Laurentide Ice Sheet 11,000 or so years before.

De Witt Clinton of Albany, later to be mayor of New York City and later still governor of the state, was the political leader of the long and acrimonious debate that followed publication of the Hercules letters. In spite of a reluctant and unimaginative response in New York City Clinton persisted and in 1811 introduced into the New York State Senate a bill proposing a commission to explore a canal route. Clinton was determined not to lose out either to the Canadians or to some upstart American Great Lakes city that was bound to blossom in the future. The construction of the original canals around the Lachine Rapids and a new system being planned were warning enough: the development of a Great Lakes link with the Gulf of St. Lawrence was only a matter of time and could only be harmful to the future interests of the towns and people of the Hudson River and New York City unless the state played its part. Clinton became fanatically determined to get New York into the act. Clinton's project was approved by the state legislature in 1816, and incredibly the Erie Canal with all its 82 locks and its 584-kilometer length was completed and operating by the end of 1825.

The main difficulty, one of many during the construction, was surmounting the Niagara escarpment at Lockport, the last hurdle before Buffalo at the eastern end of Lake Erie. The Pennines, the backbone mountains of England, had presented a similar situation in the building of a canal from Yorkshire to Lanca-shire, the Leeds and Liverpool Canal, completed in 1774. There was a fault-line escarpment within steep Yorkshire hills near the mill town of Bingley, below Charlotte Brontë's Haworth Moor, and this had been overcome by the construction of a five-chambered set of canal locks known as the Bingley Five-rise Locks. This engineering wonder enabled barges to be raised up and down a steep hill. The Erie Canal builders sent a surveyor in 1817 to study the canals of England.

Among other ideas the surveyor borrowed the Bingley Five-rise. The Erie Canal designers, who were new to the game and therefore uninhibited by conventional engineering traditions, sensibly doubled the carrying capacity of the staircase by building two sets of five-rise chambers side by side (and later regretted not going further). With the last problem solved and the last carefully dressed stone block in place Governor De Witt Clinton slung a symbolic pail of Erie lake water into New York Harbor on October 25, 1825: Clinton's Ditch was open to traffic. The Ditch was the single most important technical contribution to the future pattern of commercial and industrial development of its time.

The Erie Canal opening matched the completion of the first full-sized canal at Lachine Rapids, and it had anticipated the completion of the first Welland Canal, which joined Lake Ontario to Lake Erie on the Canadian side of the Niagara River in 1829. By 1830 there was therefore a through water route from the tidal reaches of the St. Lawrence River to Lake Michigan, via the Detroit River from Lake Erie to Huron. Significantly in 1830 a United States government engineer completed a map of the neighborhood of Fort Dearborn, where the Chicago River flowed into Lake Michigan. Apart from the fort and the desultory river there was little else to map.

The Chicago River is about 40 kilometers long and rises on the east slope of a terminal moraine deposited during the Wisconsin advance of the Laurentide Ice Sheet. The river was discovered by Louis Joliet in July, 1673, in company with a Jesuit priest, Father Jacques Marquette, during the course of their return journey from an exploratory trip down the Mississippi River from Wisconsin. On the way back north, Joliet had found a likely looking tributary of the Mississippi, the Illinois River, and had followed it up to its headwaters on the west slope of the Chicago River moraine. In his report to his patron, the governor of French Canada, he stated that an "all-water transit between Lake Erie and the Mississippi was possible." He was actually referring to Lake Michigan and the Mississippi, and he meant that it would be simple to cut a way through the moraine to join Lake Michigan to the Illinois River and thus to the Mississippi and the Gulf of Mexico. Five years later another French explorer, Sieur Robert de la Salle, saw the region and in his report stated that Joliet had proposed his ditch "without regard to its difficulties." La Salle was right about this: There proved to be enormous difficulties, as we shall see. However, la Salle was in no position to criticize: In 1667 he had been looking for China and had named a region near Montreal *la petite Chine* — which gave those fateful Lachine Rapids their name.

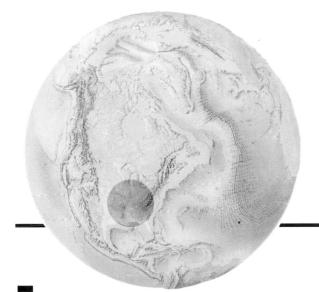

I N 1830 there were fifty settlers in the neighborhood of Fort Dearborn, nestled uncomfortably in the bend of a mosquito-infested, polluted, swamp river, the Chicago River, on the southwestern shore of Lake Michigan. Ten years later the fort was described by one observer as "a venerable relic of the comparatively olden time, in this otherwise entirely new town of Chicago." By the time of the great Chicago Fire of 1871 the city had a population of 500,000. By 1930 it had 3 million people and Al Capone. Today the Chicago metropolitan area has more than 7 million people. In North America Chicago is second in population size only to New York. Fort Dearborn's site is now the corner of Wacker Drive and Michigan Avenue, overlooked by the "corncob" buildings of Marina

City, the Sun-Times Building, and the Wrigley Building. The Chicago River now flows firmly in the opposite direction, from east to west, leading many to believe that Lake Michigan naturally drains into the Mississippi—which it doesn't.

The Illinois and Michigan Canal (the I & M) was opened in 1848, the first step in the realization of Louis Joliet's "ditch" that Robert de la Salle had been so skeptical of almost two centuries before. The I & M was cut through the barrier of compacted moraine that acted as a watershed between the Chicago River and the Illinois River, and the canal followed the route of an ancient overflow of glacial Lake Michigan, thus opening the way to the Mississippi River for vessels from the present lake. The route from the lake followed the Chicago River upstream to a lock at Summit Level, which raised traffic the modest but vital 3 meters needed to surmount the watershed, and then down through fifteen locks in 150 kilo-

meters to join the Illinois River at a town perhaps ironically named La Salle. Since the I & M system was above lake level, it had to be fed with water from another source to refill its discharged lock chambers. But the several moraine streams redirected for this purpose proved to be inadequate. Pumps were therefore installed to transfer polluted Chicago River water over the top of the hill into the locks and so out of sight and out of mind down the Illinois, 440 kilometers to the Mississippi.

An outbreak of cholera and typhoid in 1854 is reported to have reduced budding Chicago's population by 5 percent; and another in 1885, by 12 percent. The prime cause of both disasters was contamination of Lake Michigan's water by the backing up of stagnant Chicago River water. The river and the Summit Level stretch of the I & M were deepened

GATEWAY ARCH, ST. LOUIS, MISSOURI

The West (literally west of the Mississippi at the time of the Louisiana Purchase in 1803) first had to be explored; then a way had to be found to the Pacific coast. These objectives were military rather than civilian. Once tentative exploration was complete, the West then had to be settled by Americans whose previous experience of pioneering had been limited to their own backyards east of the Mississippi. They were used to a moist continental climate and not at all to an unforgivingly arid or semiarid climate. This radical change required as much adaptation of the easterners as the encounter of British-Americans with the New World had required of colonists in the 16th century; easterners suffered much hardship, and many thousands died in the process of settling the new land. Generation by generation easterners gradually adapted to the new environment and became westerners. They were a new breed of Americans.

The pioneers crossed the major barrier of the Mississippi in the vicinity of St. Louis, Missouri. They assembled a few hundred kilometers farther west at outfitting towns on the banks of the Missouri River before they began their westward trek to points sometimes several thousand kilometers or more across rivers, plains, mountains, and deserts. They traveled along the Oregon Trail and other trails in wagons, riding horseback, pushing handcarts, or simply walking. Their experiences are legendary, and their endurance incredible. Gateway Arch pictured here soars above the Mississippi River in commemoration of the pioneering phenomenon of the mid-19th century: the "western movement."

in 1871 to below lake level, thus allowing Lake Michigan's water to refill the chain of locks down to the Illinois River. The second epidemic occurred because the Chicago River (which now ambled toward Summit Level as locks beyond were discharged and refilled) had been reversed after a heavy rainstorm, polluting Lake Michigan.

As a result of a general outcry in both Chicago and "over the hill" in St. Louis the Chicago Sanitary and Ship Canal was completed in 1900, permanently reversing the flow of the Chicago River. This was the largest earth-moving operation ever attempted in North America, and the engineering advances that were made in building this canal greatly influenced the methods used in the subsequent construction of the Panama Canal (1904–1914). But the not-so-obvious result of digging the now-completed Joliet's ditch was the parallel need to develop matching technology for the treatment of polluted water — this was the real accomplishment. It was a fundamental contribution to the future of the midcontinent and a big step toward understanding the rules that govern life's survival on a stable platform.

The folks over the hill in St. Louis had built their modern town at an extraordinary intersection. Selected in 1764 as a fur-trading post, it was organized later as a village and named St. Louis after the canonized Louis IX of France. The site is just below the confluence of the Illinois and the Missouri rivers with the main north-south drainage duct of the stable platform, the Mississippi.

After his Louisiana Purchase of 1803, Thomas Jefferson decided that the continent west of the Mississippi River needed to be explored. The natural place for Captain Meriwether Lewis and Lieutenant William Clark to start their historic expedition up the Missouri (1804-1806) across the Rocky Mountains to Oregon and the Pacific was from St. Louis. The "western movement," the migration from the east to the west of the continent, followed the success of the Lewis and Clark expedition. The first wheel tracks of the future western trails were made by William H. Ashley of St. Louis (1823-1825) and his Rocky Mountain Fur Company; they headed for their Green River Rendezvous with the trappers of the Wyoming Basin over the South Pass, between the mountains of Medicine Bow and the Wind River in Wyoming. The inscription "Ashley 1825" painted on a rock in Red Canyon on the Green River took John Wesley Powell by surprise in 1869, on his way to his first journey across the Colorado Plateau through Grand Canyon.

Tens of thousands of pioneers followed Ashley's tracks across the plains to Fort Laramie and South Pass. They had to cross the Mississippi before tackling the Oregon, Santa Fe, and California trails, and the natural crossing point was at St. Louis, below the confluence. And just as naturally, the natives of the continent had also been drawn by the drainage system to this same focal point: The Cahokia of the Mississippian Culture built the largest known aboriginal conurbation on the stable platform opposite St. Louis, on the eastern bank of the river. Some archeologists estimate that the Cahokia stockaded community held more than 30,000 people at its peak in the 12th century.

The pattern for the future human development of the continent's stable platform was set by a series of tectonic events. First there was the opening of the Atlantic Ocean during the breakup of Pangea. Then came the formation of the Rocky Mountains during the Laramide Orogeny. And later, from about 30 million years ago, began the upswelling of the whole of the western part of the continent from the edge of the craton to the Mississippi basin, as a result of the subduction of the Farallon Plate. This upwarp amounted to more than 3,000 meters at its axis in the Rockies, dropping to a few hundred meters in the present Mississippi basin and to sea level in the present Gulf of Mexico. Rocky Mountains snowmelt eroded the mountains and the sedimentary rocks of the stable platform, draining generally southeastward through a system of rivers. The Mississippi basin deflected some of these rivers to the south, while the ancestral Rio Grande, Arkansas, and Red rivers made their independent ways to the Gulf of Mexico. The Laurentide Ice Sheet redirected the Missouri River from its flow to Hudson Bay and so determined the western pattern of the present Mississippi River drainage — and at the same time the mode of human development of the Mississippi basin.

The Mississippi drainage system consists of five elements: eastern and western, upper and lower, and delta. In the middle the system is constricted between the upper and lower elements like an hourglass. The eastern tributaries drain the rainy and old Southern Appalachian Mountains through well-defined valleys. The western rivers, which provide the most silt, drain the Rockies through rainshadowed, semi-arid, and easily eroded regions. The upper Mississippi River flows from the glaciated regions of the Canadian Shield, a geologically new source. The lower Mississippi has carved out a great north-south trough below the hourglass waist, the Mississippi Embayment. This section is 1,000 kilometers long, 40 kilometers wide at its northern end, and 325 kilometers wide at its southern end. The final delta is but the surface expression of a fan, a cone-shaped mound of sediment that has accumulated on the floor of the Gulf of Mexico like sand in the bottom half of the metaphorical hourglass. This fan and the sediment-filled embayment are the result of 60 million years of accumulation of river sediments. The fan alone is 500 kilometers in radius and 12 kilometers from apex to base. The part of the apex that emerges above sea level today, a fraction of the whole, is of course the Mississippi Delta. The enormous weight of material from the embayment and deltaic fan together has caused a dent, a tectonic depression called a "geosyncline," on the edge of the craton.

Geologists believe that the general upwarp of the North American continent from the line of the Mississippi River to the West Coast was caused by tectonic forces during the post-Pangean opening of the North Atlantic Ocean. Much later the weight of the Laurentide Ice Sheet depressed the central stable platform. The ice sheet's subsequent retreat left an established river system in a natural basin.

The present Mississippi drainage system resulted from the combined effect of these main events plus other events that preceded and followed them. The simplified map reproduced here shows the hourglass configuration of the Mississippi drainage system.

The basic effect of this configuration is that the largest part of the continent's inner drainage water is constricted by passage through a shallow depression above what was an ancient rifting zone, now called New Madrid after the nearby town.

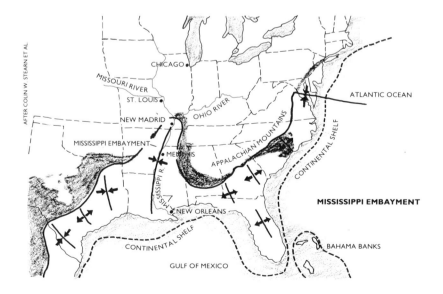

AFTER COLIN W. STEARN ET AL.

CHICAGO

MISSOURI RIVER

ST. LOUIS

OHIO RIVER

NEW MADRID

ATLANTIC OCEAN

MISSISSIPPI EMBAYMENT

MEMPHIS

APPALACHIAN MOUNTAINS

CONTINENTAL SHELF

MISSISSIPPI R.

MISSISSIPPI EMBAYMENT

NEW ORLEANS

CONTINENTAL SHELF

BAHAMA BANKS

CONTINENTAL SHELF

GULF OF MEXICO

THE MISSISSIPPI DOWNRIVER FROM NEW MADRID, MISSOURI

The Mississippi and Ohio rivers meet a short distance above the locality of New Madrid, which is pictured in this downriver aerial panorama. The stable platform of the continent lies above the confluence, and the Mississippi embayment lies below. From this point the Mississippi embayment stretches out to form a vast floodplain with a deltaic fan that surfaces in the Gulf of Mexico in the region of New Orleans.

Sedimentologists have found that a perfectly straight channel on a slightly sloping plain with unvaried river flow will form a meander if the banks are of unconsolidated and easily eroded material — as is the case of the Mississippi River. The reason for this is that river water can carry sand-sized particles only for limited distances. Sand bars form as a consequence, and these tend to deflect the river's flow. Deflection increases erosion at particular points on the river bank and in time results in the formation of sinuous bends.

Infrequent but very severe earthquakes are one of the catastrophic events that in this vicinity have caused changes in the course of the Mississippi. Some of these changes can be seen in the panorama and in more detail to the far right. According to the USGS a series of earthquakes in 1811 and 1812 is the most severe to have been cataloged on the North American continent. The illustration at left compares the effect of the New Madrid earthquakes of 1811 and 1812 with the effect of the San Francisco earthquake of 1906.

Because of the danger to the 20 million people who now live and work in the area that could be severly shaken by a future New Madrid earthquake of the same proportions, the region is being studied intensively. From these studies tectonic and seismic scientists have concluded that New Madrid is above the site of a wide graben, which itself resulted from complex continental rifting between 700 and 500 million years ago, rifting of the kind that is occurring today in the region of East Africa and the Red Sea. The illustration at the right shows minor earthquake tremors that have been recorded recently in the region. It also shows the possible outline of the underlying rift, which is estimated to be about 5 kilometers below the surface.

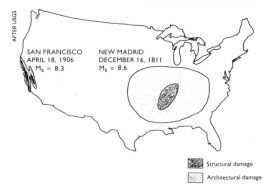

AFTER USGS

SAN FRANCISCO
APRIL 18, 1906
$M_s = 8.3$

NEW MADRID
DECEMBER 16, 1811
$M_s = 8.6$

Structural damage

Architectural damage

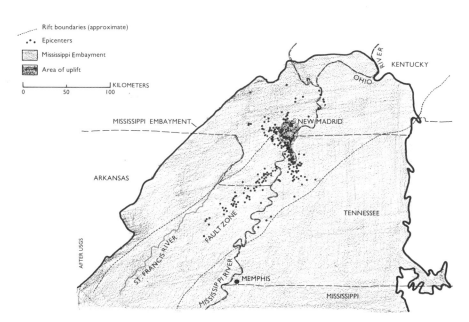

Rift boundaries (approximate)

Epicenters

Mississippi Embayment

Area of uplift

KILOMETERS
0   50   100

MISSISSIPPI EMBAYMENT

NEW MADRID

OHIO RIVER

KENTUCKY

ARKANSAS

FAULT ZONE

ST. FRANCIS RIVER

MISSISSIPPI RIVER

TENNESSEE

MEMPHIS

MISSISSIPPI

NEW MADRID, MISSOURI

The Mississippi is a clear stream at its source, among a host of lakes above Minneapolis (west of the river) and St. Paul (the head of navigation) to the east of it. The Twin Cities are the apex of an inverted triangle whose base is defined by the source of the Mississippi in Lake Itasca, 250 kilometers northwest, and the terminus of the St. Lawrence Seaway at Duluth, Lake Superior, 200 kilometers northeast. The actual length of the Mississippi from its source to the Gulf of Mexico is 3,779 kilometers although the direct distance is 2,000 kilometers. (River distance is misleading because of meanders; direct distance between points will therefore be used throughout.) But 750 kilometers south of Minneapolis, after its confluence with the silt-loaded Missouri, the Mississippi assumes its true Huckleberry Finn image. The burly river murmurs quietly but flows swiftly through St. Louis, an astonishing swiftness and ease despite its width and bulk. The river flows another 200 kilometers southward to its confluence with the main eastern drainage river, the Ohio, which drains the Southern Appalachian Mountains. And so the Mississippi flows into the waist of the hourglass.

The waist of the hourglass encompasses four state boundaries: Missouri and Arkansas west of the river and Kentucky and Tennessee to the east. This is where the lower Mississippi and its wild meanderings begin. The river's centerline determines the peculiar state boundary lines, and the state of Missouri has a very odd corner patch that protrudes into the northeastern tip of Arkansas along the line of the St. Francis and Mississippi rivers. In 1541 Hernando de Soto, the first European to explore the region, named the area "New Madrid." While surrounded by a battery of seismographs linked to monitors in the New Madrid area, I learned from Brian J. Mitchell, of St. Louis University, that there could be a geological reason for the odd-shaped piece of Missouri. Since 1974 Mitchell and his colleagues at St. Louis and those at Memphis State University and geologists of the USGS and those of other organizations have in a sense been investigating this reason.

New Madrid is at the entrance to the waist of the hourglass, and here the river flows into a shallow depression of which the odd-shaped piece of Missouri forms a part. The depression is in fact above a graben, the surface manifestation of a Precambrian rifting zone, about 5 kilometers below the surface, that was active about 650 million years ago. This is known because local samples of alkaline volcanic rock (which has been dated) have been recovered from drilling cores; such alkaline rock is commonly associated with rifting. In addition, gravity measurements that indicate higher density have shown the presence of plutonic masses in the region, and these too are deep beneath the surface. The plutons are thought to have been caused by a Precambrian hot spot in the asthenosphere and to be similar in character to plutons associated with such rifting zones today as the East African Rift Valley and the Reykjanes Ridge in Iceland.

Several interpretations have been made of the data gleaned in the last few years from the New Madrid graben. Some of the scientists consider the rift that the graben represents to have been a spreading center, perhaps one arm of a triple junction, where three plates met. The rift could prove to have been of considerable length, and some consider that it might have been associated with other continental faults. Other opinions suggest that it may prove to have been the "failed" arm of a triple junction or even that it is an incipient rift that is reactivating. Whatever the answer, the consensus is that the New Madrid rift 5 kilometers below the surface of the Mississippi River is 65 kilometers wide and 200 kilometers long and is now the axis of a seismic zone, an earthquake center.

During and since Pangean times the earthquakes in this area have caused vertical displacement totaling up to 100 meters. But more relevant to the Mississippi floodplain than displacement is the vibration caused by earthquakes. Such vibration liquefies unconsolidated material, the very substance of the lower Mississippi basin's surface. Fortunately, New Madrid episodes occur in spasms severe enough to cause major earthquakes only once every 600 years or so. But because the craton is very thick and dense beneath New Madrid, when they occur, their severity is tremendously enhanced. Their effect is much more widespread than it is in faulted coastal regions in Alaska or along the coast of California, where shock waves are dissipated by the seabed. That part of the craton with New Madrid at its center is like a huge brass gong that vibrates loud and long when struck by an earthquake; an earthquake of similar force on the coast, violent though it may be for its unfortunate victims, is a tinkling bell by comparison.

At 2 A.M. on Monday, December 16, 1811, a fleet of rough flat-bottomed boats moored on the Mississippi near New Madrid was violently upset. As in the contemporary reports of the Lituya Bay earthquake in Alaska, New Madrid earthquake survivors reported "a most tremendous noise equal to the loudest thunder, but more hollow and vibrating." The scene was chaotic: The river was displaced, riverbanks collapsed, boats overturned and sank or were impaled on "planters" (tree trunks stuck in the bottom of the river). Hundreds of people lost their lives in this first of three major shock sequences. But there was much worse to come. The following days were terrifying. In each sequence, shock succeeded shock, hour after hour, day after day. Another sequence began on January 23, 1812. A third shock on February 7 was the worst of all. In all there were literally hundreds of severe and dozens of formidable earthquakes between December, 1811, and March, 1812. A very large chunk of the central stable platform was affected. The three main shocks were by far the most severe earthquakes to have taken place in North America in historic times.

Two moderately severe earthquakes have occurred in the New Madrid region since then (in 1843 and 1895), but there have been few since of an intensity that people could feel. However, on a day-to-day basis there are almost continuous micro-shocks that stimulate the seismographs at St. Louis upriver and those at Memphis State University downriver to flick their pens to and fro on moving rolls of graph paper. The seismologists' concern, the main purpose of their many monitors at

New Madrid, is to understand the cause of the earthquakes and particularly to develop a means of predicting major earthquake trends in what is now a highly populated region: today more than 20 million people live within 600 kilometers of New Madrid.

New Madrid is just below the confluence of the Mississippi and Ohio rivers. Here the Mississippi is about 2.5 kilometers wide and still 800 kilometers due north of New Orleans; it is more than 900 kilometers from the Gulf of Mexico. Although the possibility of major earthquakes in the New Madrid region is real, it is mercifully remote; but the hazard of devastating floods is an annual event, as the presence of nearby Bird's Point Floodway testifies. The Mississippi below New Madrid and beyond to New Orleans flows across a plain on which floods are commonplace, and floods of biblical proportions occur roughly once a century. In the 20th century there has been a disproportionate number of floods of this magnitude.

Motoring down the highways of the Mississippi floodplain, I found it difficult to relate these superlative roads and the rich rolling landscapes to the need for flood control. I could well understand the naiveté of the early settlers: No dangers are apparent, and the depth and extent of the rich silt soil formed obviously suggest valuable fertility. I had read of Mississippi flood problems and had seen stories on them on television news programs. But I found that, in order to understand the scale of the flooding that the residents of the Mississippi plain have to suffer, I had to compare the size of the lower Mississippi drainage with geography more familiar to me.

I took a map of the Mississippi river system and superimposed it on a map of the British Isles. The Mississippi River at New Madrid approximately coincided with John O' Groats at the northernmost tip of Scotland, and New Orleans with central London. The Mississippi Delta covered a south-trending arc from the

CYPRESS SWAMP, LOUISIANA

The year-round humid warmth of the Gulf Coast encourages prolific growth of freshwater swamp flora, particularly the bald-cypress trees of the type pictured here. There are also countless numbers of water hyacinths, spider lilies, ferns and creepers, and other swamp plants. As freshwater environment gradually changes to delta saltwater environment (the tidal effect is small), the nature of the flora changes. Trees, for example, can no longer survive, and swamp gives way to treeless marsh. In freshwater marsh there are cattails, and in saltwater marsh there are salt grasses. Because of lush year-round growth and modest tides the change in Louisiana environments from fresh to salt water is sudden. In estuarine environments in harsh and varied climates swamps do not exist; the change from freshwater marsh to saltwater marsh is gradual.

Such watery and dependably warm, highly vegetated areas support a wide variety of animals. Louisiana swamps and marshes teem with life: crustaceans, amphibia, reptiles, mammals, insects, and birds. A greater number and variety of birds arrive and depart in the delta than anywhere else in North America. The delta is in fact the terminus of the Mississippi Flyway, the most used of the four major North American migratory-bird routes. From here each spring millions of birds of a wide variety of species fly up the Mississippi Valley to the region of New Madrid and then fan out along the Ohio Valley and the Missouri Valley. Some continue up the Mississippi to the Great Lakes, Hudson Bay, and the Arctic beyond. All such migratory species return to the Mississippi Delta as winter approaches their summer sojourns.

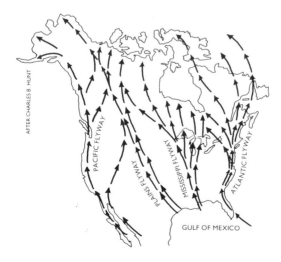

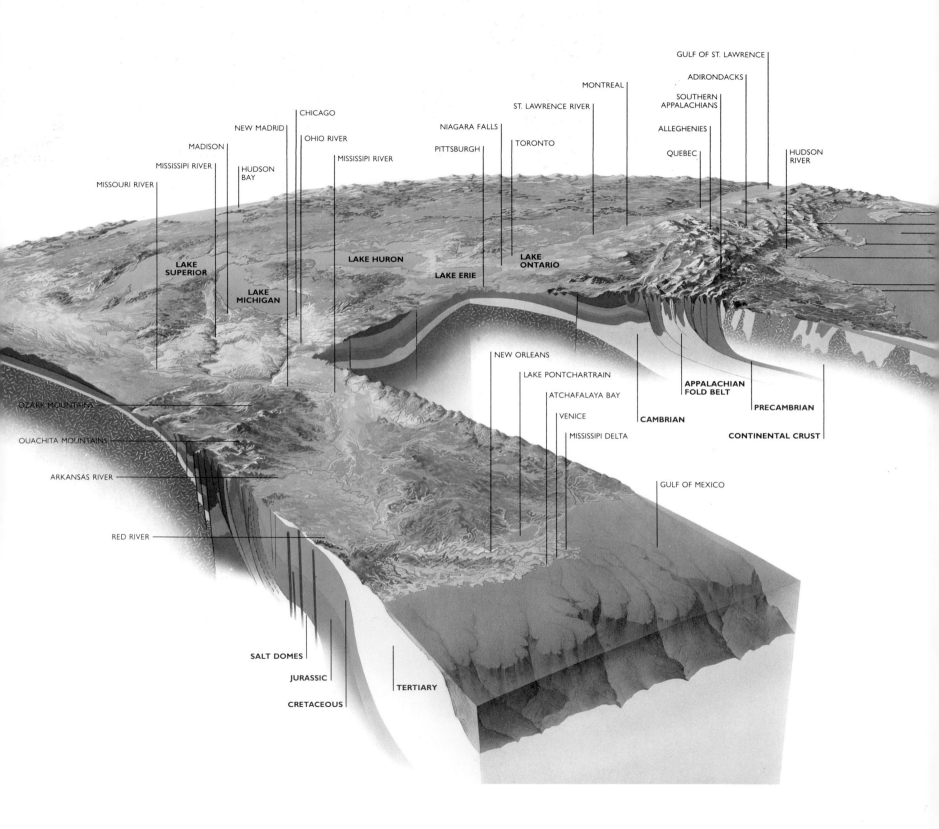

MISSOURI RIVER

MISSISSIPI RIVER

MADISON

NEW MADRID

HUDSON
BAY

CHICAGO

OHIO RIVER

MISSISSIPI RIVER

PITTSBURGH

NIAGARA FALLS

TORONTO

ST. LAWRENCE RIVER

MONTREAL

GULF OF ST. LAWRENCE

ADIRONDACKS

SOUTHERN
APPALACHIANS

ALLEGHENIES

QUEBEC

HUDSON
RIVER

**LAKE
SUPERIOR**

**LAKE
MICHIGAN**

LAKE HURON

LAKE ERIE

**LAKE
ONTARIO**

OZARK MOUNTAINS

OUACHITA MOUNTAINS

ARKANSAS RIVER

RED RIVER

NEW ORLEANS

LAKE PONTCHARTRAIN

ATCHAFALAYA BAY

VENICE

MISSISSIPI DELTA

**APPALACHIAN
FOLD BELT**

CAMBRIAN

PRECAMBRIAN

CONTINENTAL CRUST

GULF OF MEXICO

SALT DOMES

JURASSIC

CRETACEOUS

TERTIARY

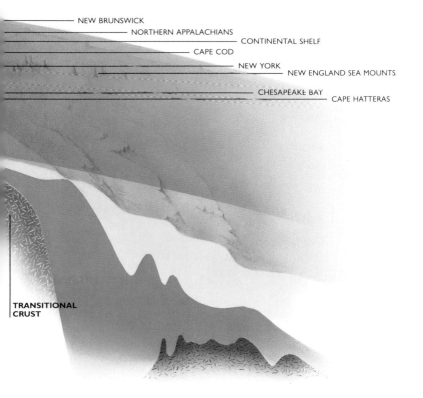

NEW BRUNSWICK

NORTHERN APPALACHIANS

CONTINENTAL SHELF

CAPE COD

NEW YORK

NEW ENGLAND SEA MOUNTS

CHESAPEAKE BAY

CAPE HATTERAS

TRANSITIONAL CRUST

This is a cutaway view of the continental regions photographed and described in this chapter and the remaining chapters of *The Making of a Continent*. The idea of the diagram was to enable both author and reader to achieve a better feeling for subsurface structures and therefore to view the present landscape with different eyes. The geography of the illustration is to scale, but the vertical scale and curvature of the Earth have been exaggerated. The result is the distorted appearance of some well-known features: the Great Lakes and the Gulf of St. Lawrence for example.

The lower block illustrates the Mississippi embayment, the fluvial plain from the New Madrid region to New Orleans. The embayment is about the same in extent as the region in Britain from John O'Groats in North Scotland to London on the River Thames. The Mississippi River sediment in the embayment extends beneath the sea in the Gulf of Mexico. The sedi-

ment has accumulated to about 15 kilometers in depth, and its great weight has depressed the basement rocks beneath it into a geosyncline. One of the consequences of such tremendous pressure and perhaps tectonic movement in the region has been to cause columns of salt from a dried-up sea beneath the present Gulf of Mexico to push toward the surface. Oil and gas migrate toward, and accumulate in, reservoir rocks near such salt domes.

The Mississippi sedimentary cover has buried evidence of possible continuity between the Ouachita Mountains west of the river (left in the illustration) and the Southern Appalachian Mountains to the east of it (shown in the upper block at the right). The upper block portrays a section through both Appalachia and the passive margin of the continent. The whole region was once the scene of geological mayhem when, among many dramatic events that preceded and followed, North America and northwest Africa collided. The region of collision between the continents can be clearly recognized from the upturned rocks called overthrust belts, which are toward the center of

the section. The smaller and extraneous-looking flat-topped rocks set into the Piedmont are suspect terrain. Some of these terrains comprise bits of "Avalonia" — a now fragmented micro-continent that is the central subject of the chapter entitled "Avalonia."

The upper block of the illustration slices through the continent from the coal-field regions of Pittsburgh, Pennsylvania, to Cape Hatteras in North Carolina (once part of Virginia). Cape Hatteras and its adjacent Pamlico Sound contain Roanoke Island, where the first British-American colonists tried unsuccessfully to establish a settlement in the late 16th century. The reason for their failure is strongly connected to local geology: The colonists tried to establish themselves on an unhealthy salt-marsh island with hazardous access from the Atlantic Ocean. Roanoke, like the rest of the coastal plain, is situated on a now-exposed but normally submarine continental shelf.

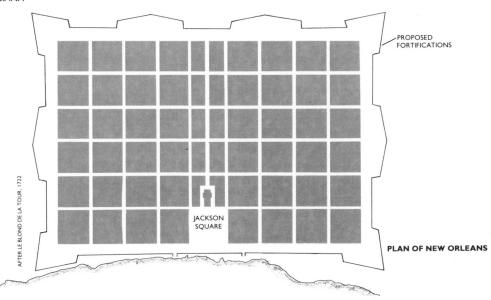

VIEUX CARRÉ, NEW ORLEANS, LOUISIANA

AFTER LE BLOND DE LA TOUR, 1722

PROPOSED
FORTIFICATIONS

JACKSON
SQUARE

PLAN OF NEW ORLEANS

The Vieux Carré, the Old Square district of New Orleans, is unique: it is built on some of the youngest geological formations on the continent; it is irretrievably sinking into unconsolidated deltaic sediments; and it is located near a "crescent" bend in the Mississippi River — a river that would now be flowing into another part of the Gulf of Mexico unless man had intervened. For all these reasons the Vieux Carré is a delightful symbol of one of mankind's major misjudgments of natural continental forces.

The pictures show buildings in Jackson Square, the central park around which the original city was built. The pictures are of the 18th-century St. Louis Cathedral to the far left and part of the Portalba Building (left, above). The Portalba Building is in two half blocks on opposite sides of the park and is the first apartment house to be erected for the purpose in America (completed in 1850).

The interior of the St. Louis Hotel is typical of many building designs in the Vieux Carré: The very hot, humid climate is excluded as much as possible from the central courtyard by sluicing its floors with water in the early mornings. The water evaporates and thus cools the enclosed, undisturbed air of the courtyard as outside temperatures increase.

People from the Old World planned and built New Orleans and settled their nearby plantations in reclaimed swamps and marshes behind natural levees. They did not realize that the Mississippi River is subject to exceptional flooding every few years, to extraordinary flooding once a decade, and to floods of biblical proportions perhaps once a century. Having established New Orleans and surrounding plantations behind natural levees, the settlers were then faced with a major threat to their property every time the Mississippi topped its banks. Their solution was to increase the height of the banks by building an artificial levee on top of the natural one.

The settlers did not realize that, to be effective, such artificial levees would need to withstand the floods that occurred "once a century." By constructing levees and then confidently expanding the region behind the leveed banks, the founders of New Orleans in Louisiana initiated a new technology — one that was first to compound their problems and then to stretch modern civil-engineering knowledge to the absolute limit of human ingenuity.

Wrought-iron decoration surrounds the balconies of most buildings in the Vieux Carré. Such work richly contributes to the ambience of the old city. Most of the beautifully executed ornamental work pictured here was made by Louisiana slaves, who were employing traditional skills.

The fashioning of figures and designs from wrought iron is an ancient African craft.

Bristol Channel to the Wash. In addition to this, a very large western river joined the Mississippi system from somewhere in the North Atlantic at Glasgow (the Arkansas River, 2,225 kilometers long) and a second major western river system from somewhere in the Atlantic west of Ireland (the Atchafalaya-Red system of rivers, 1,950 kilometers). Almost the whole of Scotland, Wales, and England was required to contain the Mississippi floodplain.

The early Europeans who colonized the Mississippi Embayment did not realize that flooding is the means by which rivers build up fluvial plains and deltas. In fact the decision to build a town on the Mississippi Delta was made in Paris in 1717 by a Scottish banker, John Law, whose Company of the West had been given control of Louisiana by the French regent, the Duke of Orléans. The man who selected the site was Jean-Baptiste de Moyne, Sieur de Bienville. In March, 1718, Bienville supervised the clearing of swamps on the banks of the Mississippi River south of Lake Pontchartrain. What is believed to be the first formal plan for the city was produced by Le Blond de la Tour in 1722.

Unfortunately Bienville had selected a site that later proved to be the apex of the cone of unconsolidated sediment. Meanwhile settlers in New Orleans and plantation owners spread along nearby river banks had established themselves safely behind natural levees, not realizing that a natural levee is no more than the insecure edge of a river bank raised by the accumulated silt of the last high flood. The founding fathers of New Orleans and their successors left a marvelous cultural mix, of which the Vieux Carré is the heart. But they also left a fearful legacy for today. The miracle is that for 250 years a combination of tenacity, courage, and technology, driven by the sheer need to survive in this location against all odds, appears to have gained temporary mastery over the Mississippi. But the plain fact is that the Mississippi, like the few other rivers on that scale, is its own master and will ultimately govern the destiny of its fluvial plain whatever is done to prevent it.

The Natchez Indians of the Mississippi floodplain built their dwellings on stilts. If

During flood the Mississippi River distributes silt to form a silty-clay surface on its fluvial plain. At "normal" river level the Mississippi takes the easiest slope seaward and then cuts a sinuous channel in its clay bed. At flood level, when the river channel is brimful, the river deposits extra silt and clay on its banks. This deposit accumulates from repeated inundation to form what is called a "natural levee." The Mississippi therefore builds its own banks above the level of its floodplain.

AFTER NEW ORLEANS GEOLOGICAL SOC.

NATURAL LEVEE

SWAMP · BRACKISH MARSH

CHANNEL FILL

SUBSIDENCE

OLD CHANNEL FILL

SUBSIDING LEVEES CAUSE SWAMP FORMATION

This plantation house, now restored from decay, is characteristic of the American antebellum period. Houmas House is on the Mississippi River above New Orleans. It was built in the early 19th century behind what was then a natural levee. Houmas plantation had 20,000 acres of sugar cane and four mills to process the cane crop. The house became the headquarters of one of the largest plantations in Louisiana. These were good reasons for the owner to worry about the height of natural levees.

During its checkered history the Houmas plantation employed several thousand slaves, housed on the plantation — and it was differences about such employment in this and many other plantations (producing tobacco and cotton as well as sugar) that precipitated the Civil War of 1861 to 1865. Houmas House and grounds are now devoid of both sugar cane and processing mills: They have been replaced in recent times by innumerable oil refineries downstream.

The trees that frame the picture have strands of Spanish moss hanging from them. The hair-like moss (supposedly resembling a Spanish beard) is an epiphyte, an air plant, which grows on host trees in cypress swamps and beside rivers in subtropical climates. The propagation and cultivation of Spanish moss was once an extensive industry. The dried material was mostly used for the upholstering of mattresses and furniture. Up to a ton a year of fresh moss was harvested from large trees in shady swamp forests.

floodwater looked too menacing, they sought higher ground and stayed there until the flood receded. Although they tried other methods, in the main the newly arrived Europeans also built their houses on stilts. They drove wooden piles into the swampy ground to support structures and in fact still use them today even for the tallest and most modern buildings — except that modern piles are of concrete driven in by pile drivers. The Europeans were not prepared to abandon their homes and plantations during floods; so they built artificial levees about 1 meter high on river banks near New Orleans. They soon found it necessary to extend these levees considerably.

The river slowly became a prisoner of its leveed banks: It could no longer disperse its silt and water. It rose to higher levels between levees during periods of normal flooding; the river's discharge was confined to a particular point on the delta. The river mouth became artificially elongated and had to be dredged constantly. Raising the height of levees and dredging the river mouth channel became a preoccupation. Three-meter levees were now necessary along considerable distances of river bank. Battling the Mississippi became a major political issue. In 1879 a River Commission was established, and in 1881 the U.S. Army

Corps of Engineers were made permanently responsible for engineering and maintenance on the river. No sooner had this important step been taken than a disastrous flood in 1882 overwhelmed the now 3-meter levees. By 1926 even more extensive and even higher levees and dams had been built at enormous cost. It was thought that at long last the war against the river had been won. But weather trends in the autumn of that year and abnormal river levels in the regions above New Madrid through the early months of 1927 were ominous. Through March the rains deluged much of the central stable platform, and during the night of April 15 an incredible 36 centimeters of rain fell on New Orleans. A cloudburst in southern Kansas caused the Arkansas River to burst a levee, and the first of the biblical floods to be experienced by modern man was terrifyingly on its way. The sad lesson learned was that, whenever a levee was breached upstream, the danger to New Orleans was lessened. There were many such breaches in the levees in the days immediately following April 27, 1927.

Frederic M. Chatry, Chief of the Engineer-

HOUMAS HOUSE, NEW ORLEANS, LOUISIANA

ing Division of the Corps, explained the change in policy that followed the 1927 disaster by suggesting to me, when I visited his office in New Orleans, that on the night of April 27, 1927, the war with the Mississippi had been lost. A new mood of enlightened compromise with the river had followed. Instead of just *containing* the river, the Corps built floodways that directed floodwater down leveed routes such as Bird's Point Floodway at New Madrid; reservoirs were created that can be filled at peak and discharged at low water; cutoffs were provided with a complex system of internal routing to alleviate threatened areas — such is the present use of Lake Pontchartrain. The Corps has also developed an array of special equipment: bank-grading and concrete mat-laying plants; fuse-plug levees and replaceable wooden "needles"; suction-dredging fleets; pumping stations; and in the jargon of the river "ripraps, wing dams, and contraction dikes" to control day-to-day river flow and silt precipitation. Most important, a code of navigation was established. The key to the safety of the levees is the regulation that governs wash height; the owner of a vessel pays the cost of repair for any damage caused. In addition the levees were yet again made higher, this time to 10 meters (450 square meters in section) compared with the 3-meter maximum (25 square meters in section) of 1882.

The aim was and still is to establish a defense against the worst flood that geologists can infer from the sedimentary record. In the special language of the Corps, such a doomsday flood is called the "project" flood level. The first real test of the system was in the 1937 flood, "one of the greatest of all recorded floods." But there was another in 1973. And yet another is in progress as I write these words in 1983. The spillways are deliberately flooded during such crises and are often depicted in newsreels as disaster areas. So they are for the unfortunate people who have to abandon their homes. But such residents of the spillways know the risks and are obliged to sign easements to this effect if they choose to farm the very fertile floodplain. There were anxious and even desperate moments of crisis during the peak flood years of 1937 and 1973,

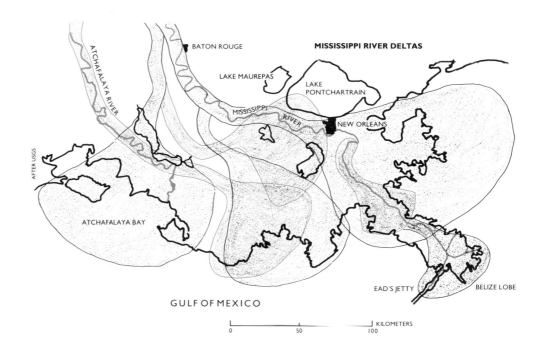

During the last 5,000 years the Mississippi River has changed its point of discharge into the Gulf of Mexico a number of times. This drawing shows the seven main subdelta lobes that have formed during this period. The two lobes that are discussed in the narrative are the one that terminates in Atchafalaya Bay at the left of the drawing and the mostly artificial lobe (the Belize) toward the right-hand corner of the drawing.

In this panorama the Vieux Carré, the original site of New Orleans, is straight ahead on the bend in the Mississippi River. Because of this bend the city is named "The Crescent City," but because of New Orleans' geological location it is a city with extraordinary problems. As the deltaic sedimentary pile on which it is built is consolidating, the whole region is slowly sinking.

The banks of the river are artificially but very formidably leveed, and an additional levee system also segments New Orleans itself. Although the present height of levees and the use of emergency floodways on the Mississippi embayment above New Orleans have coped with recent very high floodwater levels, no one really knows what the future holds. The future weather patterns on the continental stable platform above New Madrid and elsewhere determine the floodwater level of the Mississippi drainage.

Most past and present construction of buildings and bridges in New Orleans makes use of piles, which are driven into the sediments. A century or more ago cypress trees were turned to this purpose; for they had to be cut to clear the swamps that covered the site of the future city. The Pontalba Building, for instance, is underlain with a cypress grillage; the building has shown little sign of settling since it was completed in 1850. On the other hand the modern Mississippi River Bridge at the foot of this picture has suffered from settling, in spite of the fact that a great number of piles had been driven 18.0 meters or more into the clay and spread over 6.5 kilometers of bridge approach on either side of the river.

NEW ORLEANS, LOUISIANA

AFTER U.S. ARMY CORPS OF ENGINEERS

RED RIVER

RED RIVER BACKWATER AREA

OLD RIVER CONTROL

MISSISSIPPI
LOUISIANA

FUSE PLUG LEVEE

ATCHAFALAYA FLOODWAY

MORGANZA FLOODWAY

GUIDE LEVEE

GUIDE LEVEE

MISSISSIPPI

BATON ROUGE

LAKE PONTCHARTRAIN

ATCHAFALAYA BASIN FLOODWAY

NEW ORLEANS

BONNET CARRE SPILLWAY

RIVER

LOWER ATCHAFALAYA RIVER

ATCHAFALAYA BAY

GULF OF MEXICO

VENICE

KILOMETERS
0   50

EAD'S JETTY

MISSISSIPPI RIVER FLOOD CONTROL SYSTEM

The Mississippi River is unstoppable. It will continue to flow, to flood, and to deposit sediment probably for millions of years into the future. At the moment the river is being contained and directed mainly where man wants it to go. The result is that the artificially extended Belize sublobe of the Mississippi Delta is getting longer and more difficult to dredge and to maintain.

If the river had its own way, it would by now be discharging into Atchafalaya Bay instead of being

restricted by the Old River Control Structure shown in the map.
If the river had *not* been controlled by engineering, essential freshwater supply to industry on the banks of the Mississippi near New Orleans would have stopped.

Some geologists think that Old River Structure control of the Mississippi can only compound difficulties for future generations. They suggest that a study should be made *now* of the effect of allowing the Mississippi to take its natural course while maintaining the industrial viability of New Orleans.

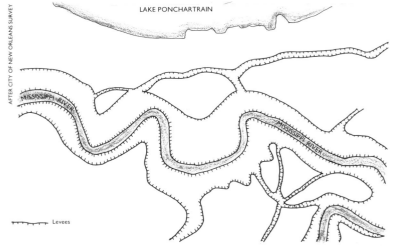

LAKE PONCHARTRAIN

AFTER CITY OF NEW ORLEANS SURVEY

MISSISSIPPI RIVER

MISSISSIPPI RIVER

⊢⊢⊢⊢⊢ Levees

NEW ORLEANS METROPOLITAN LEVEE SYSTEM
Regions inside levees are below normal river level

MINNEAPOLIS

SIOUX CITY

CHICAGO

PITTSBURGH

MISSOURI RIVER

ILLINOIS RIVER

OHIO RIVER

ST. LOUIS

ARKANSAS

CUMBERLAND R.

CAIRO

WHITE R.

TENNESSEE R.

RIVER

MEMPHIS

HUNTSVILLE

RED RIVER

OUACHITA R.

MISSISSIPPI RIVER

YAZOO R.

TOMBIGBEE R.

BLACK WARRIOR R.

ALABAMA R.

VICKSBURG

BATON ROUGE

HOUSTON

NEW ORLEANS

GULF OF MEXICO

MISSISSIPPI RIVER NAVIGATION SYSTEM

THE NATCHEZ

The *Natchez* on the Mississippi River at New Orleans is a modern steel reproduction of the original wooden boat that raced the *Robert E. Lee* from New Orleans to St. Louis in 1870. When this picture was taken, the *Natchez* had just returned from an annual commemorative race.

From the early years of the 19th century until 1865 steamboats had a monopoly of transport throughout the navigable waters of the Mississippi drainage (see map). At first they were welcomed to the towns and villages on the banks of "western" rivers. Steamboat crews announced their arrival from afar by loudly playing popular tunes on the calliope, a steam organ perched on the top deck of the steamboat. But later, as Mark Twain recounts in his stories, steamboats proved to be not only a means of bringing entertainment and freight to developing towns but also a means of distributing doubtful characters throughout communities along river banks. In time riverboats became unwelcome visitors.

After the establishment of railroads the steamboat era began to wane, and until 1925 the boats were used only as a convenience. Today they still have some economic advantage, but they are used mainly for nostalgic pleasure. But in 1885, as the Currier and Ives panorama reproduced here demonstrates, New Orleans, like many other riverside cities on the network, was a thriving hive of steamboat activity.

THE NATCHEZ'S CALLIOPE

NEW ORLEANS IN 1885, ACCORDING TO CURRIER AND IVES.

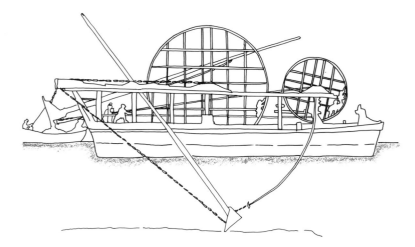

The Mississippi River is a huge, swift, silent, silt-laden, and debris-filled river with a will of its own. Nevertheless, any insoluble ingredient from sand-sized grains upward tends to precipitate in eddies and at bends in the river. Keeping navigation channels open for vessels therefore requires continuous and considerable effort.

The type of dredge shown here is an 18th-century French concept (courtesy of the Memphis District Corps of Engineers). Present-day dredges are floating power plants that use a suction pump and dustpan technique on the river bottom. The "vacuumed" sediment is discharged out of harm's way by pumping it through a long outboard pipe supported above the river surface on floats.

The Mississippi River has dictated the invention of a large number of strange-looking utilitarian craft, of which this is one: Henry Shreve's snag boat of 1838. It was used to remove tree stumps stuck in the silty river bottom and likely to snag boats and cause them to sink.

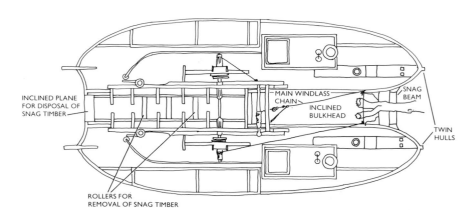

INCLINED PLANE FOR DISPOSAL OF SNAG TIMBER

MAIN WINDLASS CHAIN

INCLINED BULKHEAD

SNAG BEAM

TWIN HULLS

ROLLERS FOR REMOVAL OF SNAG TIMBER

but the system from New Madrid to New Orleans worked. However, Chatry was careful not to say that the Engineers are satisfied: I am quite sure that he did not want to tempt fate.

And what of the future? I visited a consulting geologist, Charles R. Kolb, of Louisiana State University, who told me the story of the steamboat captain Henry M. Shreve, who in 1831 shortened his route of navigation from the Mississippi River to the Atchafalaya River (and the Red River system) by cutting through a narrow neck between the rivers' meanders about 200 kilometers northwest of New Orleans. Unfortunately the Atchafalaya provided a quicker, shorter, and slightly steeper route of drainage for the Mississippi River to the Gulf. By 1950 the Atchafalaya was threatening to capture the great river through Shreve's channel, which is called the Old River. If the threatened capture had been allowed, the great port of New Orleans would have become a saltwater estuary, and the industrial belt along the river's banks that is dependent on huge volumes of fresh water would have come to a grinding halt. A parallel Mississippian system of levees and flood control had therefore to be built by the Corps to separate the Mississippi River from the Atchafalaya basin. But Kolb's point to me was that the Atchafalaya River (which now takes 30 percent of the Mississippi flow) would have naturally captured the Mississippi River without Captain Shreve's help.

From a geological viewpoint Kolb believes that in time the Mississippi is bound to have its own way. What we see today, Kolb suggests, is no more than the surface of the cone of sediment. That cone is slowly consolidating, and as it consolidates, it will sink below sea level; for the surface is no longer being replenished. Kolb advocates a study to determine the effect and method of allowing the Mississippi to divert to the Atchafalaya basin, while maintaining the industrial viability of New Orleans. The danger, in Kolb's opinion, is that in spite of the great works of the Corps of Engineers the river might divert of its own accord — suddenly and disastrously.

Captain Shreve was in fact responsible for ventures rather more constructive than opening up the Mississippi connection to the Atchafalaya River. In 1827, for example, he proposed that a steam-powered "snagboat" be constructed to remove snagged tree trunks from the river bottom; for such hazards were a very great and frequent danger to riverboats: The Mississippi River today is no less dangerous for small craft. "Planters" snag debris and project above the surface, where they can be avoided only by observant river pilots. Such tree trunks that saw up and down above and below the surface are called "sawyers." My guess is that a one-time Mississippi river pilot turned writer, Samuel Langhorne Clemens, named a fictitious character Tom Sawyer after this river term. As I made my way back to my hotel in New Orleans after meeting with Charles Kolb, I meditated in a traffic jam in Canal Street, waiting to turn left into Chartres Street in the Vieux Carré, and wondered just what the writer Clemens as Mark Twain would have made of all this geologizing in his day, an age of steamboats and of human flotsam and jetsam.

People who spend their lives on or about the Mississippi River have a quality that I had heard about but had not experienced, the celebrated mixture of friendliness, generosity, and charm known as "southern hospitality. This was now exemplified for both Joy and me by Captain Clarke C. Hawley and his men aboard the modern stern-wheeler *Natchez*, a steel-hulled steamboat that sails the Mississippi from the river wharf at Jackson Square in the Vieux Carré to give many thousands of visitors like us a pleasurable experience reminiscent of the days when such boats ruled the river. Captain Hawley's knowledge of boats and the Mississippi River navigation system is legendary, and I gleaned much about river lore and steamboats from him.

It was not obvious to me until Hawley emphasized the point that in the days of sail the shallow, winding Mississippi River was especially difficult to navigate for ships from the Gulf of Mexico and thence up the river. Perhaps the original New Orleans site was chosen by Bienville as much for its proximity to Lake Ponchartrain (then readily accessible to sailing ships) as for the crescent-shaped river frontage of the Vieux Carré. To the French the region of Louisiana was a new frontier. And through the checkered years of its history of secret sale to the Spanish and then back to the French and

finally to the Americans in 1803, Louisiana remained a troubled western outpost. European colonists east of the Mississippi River thought of it as the very far western frontier, and only a few of them considered it as a possible future means of access to the center of the continent from the Gulf of Mexico. Even today many easterners talk in terms of the American West's starting at the Mississippi, when the Rocky Mountains, 1,000 kilometers nearer the Pacific, would be the more accurate demarcation.

The use of steam as a source of power to produce rotary motion came to America after James Watt's invention of the "sun-and-planet" gear in 1781. Early experiments with steamboats that followed Watt's invention were on eastern rivers. Robert R. Livingstone, after his retirement from politics as one of the leading Louisiana Purchase negotiators, obtained a twenty-year monopoly for steamboat navigation in the state of New York and formed a partnership with the inventor-designer Robert Fulton to plan the building of a steamboat. Their first, the *Clermont*, was tried out in 1807, and a regular steamboat service along the Hudson River between the city of New York and Albany was in operation soon afterward. The partnership then expanded its commercial objectives to the Mississippi River, after securing an exclusive right for steamboat navigation on the river for a period of eighteen years. This move was viewed by the river people as a bit of eastern skulduggery; they were infuriated and refused to be intimidated. By 1819 a number of groups were operating steamboats on "western" rivers, including the indomitable Captain Shreve, who in 1817 led the general defiance of the monopoly by taking the steamboat *Washington* from New Orleans up the river to Louisville, Kentucky, in twenty-five days. Shreve and others were in competition with Fulton and Livingstone, and they operated steamboats designed by John Fitch, who in fact had produced serviceable steamboats before Fulton in 1791. The colorful and extraordinary age of the steamboat had arrived on the Mississippi.

VENICE, LOUISIANA

The town of Venice, pictured here, is the first sizable town on the delta, 50 kilometers upstream from Eads Jetty. The name Venice is indeed apt: The town is built on some of the youngest sedimentary formations in North America.

As I sat in the pilot house of the *Natchez*, which was just back from an annual steamboat race upriver, Captain Hawley told me that in addition to its novelty what had caught the public's imagination in the 19th century was the steamboat's speed. In 1825 the *Tecumseh* made the upriver trip from New Orleans to St. Louis in eighteen days, and in 1849 the *Sultana* did it in five days, thirteen hours. This improved speed was found to be as dependent on the length of the steamboats as it was on their power. Along with increased length the hulls were made shallower to accommodate shallow water. The height of the superstructure was increased to four decks for maximum cargo; "gingerbread" was added for decoration; and a balanced rudder rounded off the development of the familiar "Ol' Man River" showboat design.

The steamboat was the easiest means of transporting freight and people between communities in the Mississippi basin. From a distance the showboats announced progress by means of a calliope, a steam-operated organ, arousing excitement ahead of their arrival. Steamboats provided entertainment and news to burgeoning villages, towns, and cities along the river's banks and were glamorous by virtue of their contrast with the often grim realities of life on the tobacco, cotton, and sugar plantations along the river, more often than not a happy contrast to the dreary sameness of industrial towns. The stern-wheelers were put to every conceivable use, but their main purpose was trade. As Mark Twain depicted in his stories, they were also attractive to a wide variety of ne'er-do-wells, cardsharps, and cutthroats, who were delivered along with the freight. As a consequence, by the 1850s the showboats had fallen into disrepute and were no longer welcome. But they continued to haul cotton, sugar, tobacco, and other commerce until about 1870, when their use was superseded by more efficient but less elegant tugboats and barges. Today such barges haul — or, more accurately, push — millions of tons of cargo up and down the Mississippi and its vast river system each year: Add oil to the list, and they would be little different from John Masefield's *Cargoes* ". . . of Tyne coal, road-rail, pig-lead, pinewood, iron-ware, and cheap tin trays."

THE MOUTH OF THE MISSISSIPPI

This is the present mouth of the Mississippi River on the Belize subdelta lobe in the Gulf of Mexico. It has been artificially extended by levees (Eads Jetty) built on either side to allow a channel to be dredged for ocean-going vessels making their way upstream. At this point the river mouth is 150 kilometers from New Orleans.

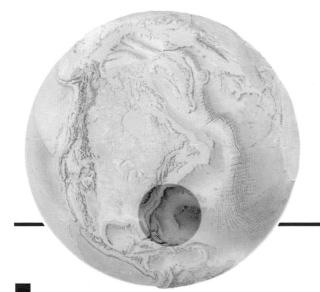

PASSIVE MARGIN

IN stark contrast with today's strings of powerfully driven barges on the Mississippi River, with their heavy industrial cargoes, 16th-century Spanish treasure galleons returning home from the Isthmus of Panama were crammed

With a cargo of diamonds,
Emeralds, amethysts,
Topazes and cinnamon, and gold moidores.

The galleons charted a northeasterly route across the Caribbean Sea. They turned north through the Greater Antilles by Hispaniola, and after threading their way through the Bahamas near San Salvador Island, they crossed the Tropic of Cancer, the latitude that marks the edge of the tropics. The ships were on their way to a point near Madeira and the Canary Islands, on a course that would take them to the Strait of Gibraltar and the Mediterranean. The familiar stanza of Masefield's poem begins

Stately Spanish galleon coming from the Isthmus,
Dipping through the Tropics by the palm-green shores,...

When John Masefield composed "Cargoes," he not only wrote a memorable poem but unwittingly provided an almost perfect description of the path of the late-Pangean Tethys Seaway.

Before the breakup of Pangea into the present continents, which began in earnest around 140 million years ago, there was no Atlantic Ocean. There were no ice caps at the poles or anywhere: Sea level was therefore at least 50 meters higher than it is today and became higher still as the Mid-Atlantic Rift began to function and Cretaceous seas spilled onto continental stable platforms. Pangea began to rift into the two continents of Laurasia to the north and Gondwanaland to the south, and as it did so, an east-west sea opened between the two masses, the Tethys Seaway. The Tethys Seaway was equatorial and in places shallow and so warm, soup-like, and salty that beneath the surface little if any free oxygen could remain in suspension. Yet in these shallow parts the seaway was teeming with minute organisms, which, having lived their brief lives, sank to the bottom and accumulated in unimaginable numbers. In the millions of years that followed, such organic matter was intermixed with, and then covered by, sediment. As the thickness of sediment increased, the effect of pressure and so of heat upon the organic con-

OIL-DRILLING AND WELLHEAD PLATFORMS, GULF OF MEXICO

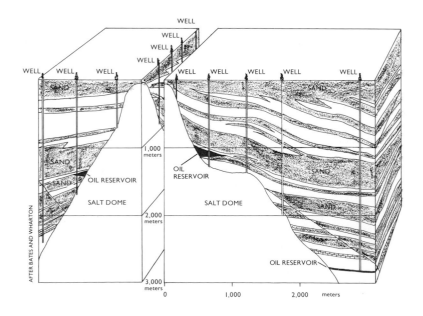

AFTER BATES AND WHARTON

Relatively small, near-surface, manifestations of salt domes measure up to 10 kilometers in width, but their subsurface structures are huge. There are hundreds of such domes in the Gulf-coast region, and they vary considerably in shape and size. The most accessible are those on shore in the area of the Mississippi Delta (see diagram at right) and its floodplain. Salt domes also occur in less accessible places beneath the sea off the Gulf coast.

All the oil platforms shown in the panorama above are perched above salt structures and are either drilling wells or drawing oil from reservoirs of oil or gas that have accumulated in rocks on the flanks of such domes.

The formation of salt domes is a continuing process. The weight of fresh Mississippian sediment on top of old

In the Gulf of Mexico oil-drilling and wellhead platforms are figuratively the modern counterparts of the stately Spanish galleons of the 16th century. The Sun's energy was the common factor in the formation of the oil now brought to the surface at wellheads and in the source of wind power for filling billowing galleon sails.

These platforms are modest affairs built in shallow water. Deep-water platforms, such as those in the North Sea off North Scotland, are about the height of the Eiffel Tower. A new type of platform now being planned for use in the Gulf of Mexico will be as tall as the World Trade Center towers in New York.

In the Gulf of Mexico there is a special hazard for oil rigs. Unlike "Hereford, Hertford and Hampstead where hurricanes hardly happen" hurricanes are devastating occurrences off the Gulf coast. Masses of humid tropical and subtropical air are influenced by gyrating sea currents and by humid Atlantic air. The Sun's energy is stored in the form of supersaturated warm air that gradually takes the shape of a counterrotating disk — anti-cyclonic hurricanes. In the Southern Hemisphere such disks rotate in a clockwise direction — cyclonic typhoons. The disk-shaped hurricane (and typhoon) has its own violent and unpredictable momentum and is fortunately a rare phenomenon.

lithified sediments builds enormous pressure on the bedrock of limestone and the sea of rock salt beneath it. Under such pressure the salt behaves like a liquid. Weaknesses in the limestone basement are penetrated by the salt, allowing a column to push up toward the surface until the pressure from beneath and from above is equalized. During this upheaval rock layers and

sediments in the path of the ascending salt are turned aside. Hydrocarbons that have formed in the sediments then migrate upslope toward the salt dome, where they accumulate in pockets of gas-filled and oil-filled reservoir rocks.

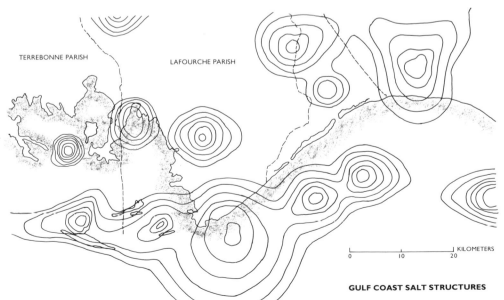

TERREBONNE PARISH

LAFOURCHE PARISH

KILOMETERS
0   10   20

GULF COAST SALT STRUCTURES

tent converted it to oil and gas. It is no coincidence that the Persian Gulf, the Aral Sea, the Caspian Sea, the Black Sea, the Balkan Peninsula, the coast of North Africa in the Mediterranean Sea, the coast of northwest Africa, the North Atlantic seaboard of South America, and the Gulf of Mexico are all prolific sources of oil. These areas are now either sedimentary basins or continental shelves, but they once formed part of the Tethys Seaway.

The Mississippi, the principal river flowing into the Gulf of Mexico, has formed a delta that is more than 12 kilometers thick and 500 kilometers in radius. Its accumulated sediment is so heavy that it has depressed the limestone basement rocks on which it rests, and has caused an immense southward-tilting, saucer-shaped depression in the craton and adjoining seabed. The saucer has a complex northern limit defined on the surface by the edge of the Mississippi embayment, with New Madrid at its focal point. The base of the cone of Mississippi deltaic sediment in the region of New Orleans extends to an ill-defined southern limit, the deeply submerged edge of the saucer beneath the Gulf of Mexico. This Mississippi

geosyncline is so vast that it appears to have actually interrupted and buried what might be related ranges of mountains, part of the Ouachita Mountains to the west and part of the Southern Appalachian Mountains.

Above sea level Pangean limestone rocks form the edges of the saucer, and on continental shelves in the Gulf of Mexico the formation of limestone has continued beneath the sea. Pangean limestone rocks accumulated in a thick platform that now contains the saucer from the edges of the Mississippi geosyncline to and along the eastern continental shelf. What one can see today of Florida (the Peninsula Arch) and the Bahama Islands (the Bahama Banks) are yesterday's limestone formations superimposed on that platform.

The weight of silt carried down to the Gulf of Mexico by the Mississippi River and accumulated on the deltaic fan has done more than depress the continental craton and the seabed: It has caused the formation of salt domes to some extent resembling Upheaval Dome in the Paradox Basin on the Colorado Plateau. In late Pangean times a basin at the western end of the Tethys Seaway evaporated, was covered with limestone, and is now a sea of salt below the basement rock under the Mississippi floodplain and delta. The tremendous pressure of Mississippi sediment pressing on this

limestone basement causes the salt below to behave like a fluid, seeking any minute crack to escape. If salt finds a weakness, it penetrates the basement and then, in the form of a column, the overlying sediments. The crystals of halite (common salt) that form in the process are from 1 to 2 centimeters in diameter and sometimes nearly 40 centimeters long. More than 300 such salt structures are known to exist in the coastal provinces of Alabama, Mississippi, Louisiana, and Texas and on the adjacent continental shelf.

Salt domes, commonly round, can be up to 10 kilometers in diameter near the surface. Sometimes the domes break the surface; and sometimes large areas of the basement limestone and other rock, called "cap rock," are raised to the surface as the salt column forces its way upward. And as the domes push upward, they disturb layers of sedimentary rock or muds or sands, which are then bent upward by the intruding column. Gradually the oil and gas that formed from the micro-organisms of the Tethys Seaway and from more recent organic material incorporated in the pile migrate up inclined sedimentary slopes toward the salt columns, where they are trapped by faults and accumulate to form reservoirs.

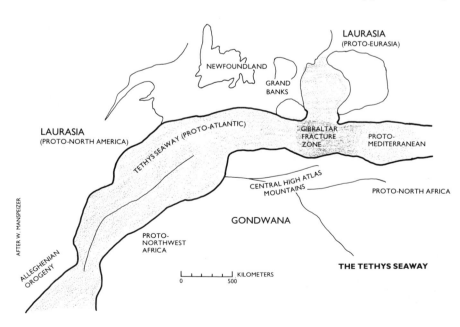

AFTER W. MANSPEIZER

The Tethys Seaway formed between the two main blocks of continents as Pangea began to rift into a northern part (Laurasia) and a southern part (Gondwana). The global map shows a general reconstruction of the geography of the Earth about 140 million years ago. The more detailed map is a close-up of an earlier stage of rifting in areas that later became the Mediterranean Sea and the North Atlantic Ocean.

The Tethys Seaway had branches and side basins. One of the main subsidiary regions of the Tethys Seaway is called the Paratethys. The present Aral, Caspian, and Black seas are remnants of the Paratethys.

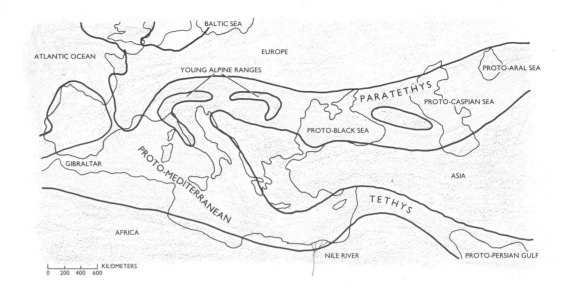

I was curious to know how salt from the Tethys Seaway could have evaporated to form a sea of salt beneath the Gulf of Mexico and how limestone could have then been precipitated on top of it. A petroleum-geologist friend, Richard W. Boebel, of New Orleans, suggested that I should look at the literature on another part of the original Tethys Seaway, the Mediterranean Sea. Most geologists, I understood, now believe that the Mediterranean's recent geological history is a good model for considering the origin of the salt basins in the Gulf of Mexico. Boebel later sent me a stack of papers on the formation of salt structures and on the exploration for oil and gas and even a primer on oil and gas production. But most importantly the stack contained notes on the formation of the Mediterranean Sea.

During the separation of Pangea into the present continents the North American continent rotated away from Eurasia. But by about 20 million years ago the continent of Africa had moved northward toward Eurasia, restricting the remaining eastern section of the Tethys Seaway to the present region of the Mediterranean and the Persian Gulf and isolating the Paratethys Sea (the interlinked Aral, Caspian, and Black seas). By about 6.5 million years ago the Mediterranean Sea was almost closed off from both the North Atlantic Ocean and the Paratethys — and at times perhaps completely so. A relatively cool and dry climate had replaced the heat and humidity of the once tropical Tethys Seaway, and because of dry air, perhaps similar to the warm dry air of the Great Salt Lake region today, the Mediterranean Sea evaporated, becoming a sea of salt that was replenished by salt water from

the Atlantic at a rate about equal to the rate of its evaporation. The result was a basin containing saturated salt water with up to 2,000 meters of salt deposits beneath it. According to some geologists the huge volume of salt that accumulated in the Mediterranean basin would require the evaporation of forty times the volume of the present Mediterranean Sea. Such evaporation appears to have taken place within a period of a few million years because the present Strait of Gibraltar opened up and allowed the Atlantic Ocean free access to the Mediterranean basin about 5 million years ago. It is probable that, in similar fashion in Pangean times, salt from the Tethys Sea produced a sea of salt beneath the Gulf of Mexico.

But how, I wondered, did such a salt basin become covered with a thick layer of limestone? This was my next objective: to see a present-day example of one of the processes that led to the formation of such limestone — a process that has played an important part in the building of the stable platform of the continent. Consequently Joy and I made our way 1,400 kilometers southeast to the "palm-green shores" of what was once part of the southern shore of the Tethys Seaway, to those cusps of coral called the Bahama Islands, which are mounted in golden-white sands set in an emerald, topaz, and sapphire-blue sea.

At the break of dawn one morning we took off from Nassau, the capital of the Bahamas, in a twin-engined high-winged aircraft. We expected to fly well over 1,000 kilometers that day on a rather carefully planned elliptical route. The farthest point was to be the claimed site of Christopher Columbus's landfall, San Salvador Island. This island is 310 kilometers south-southeast of Nassau, but I knew that our circuitous route, plus maneuvering for photography, could easily quadruple the normal two-hour direct flying time. There were no dependable refueling points en route and no seaplanes available. We settled for a slow and somewhat noisy ten-seater, an affable and able pilot, and a knockdown charter fee from a Bahamian airline. The result was without doubt one of the most colorful flights of a lifetime.

How the iridescent upper regions of the Bahama Banks were formed is well known to geologists, but whether they formed on continental or oceanic rock is a matter of debate. Most experts agree that Charles Darwin might well have been right when he suggested that tropical islands, such as the Bahamas, formed when corals and shellfish built their carbonate structures in reefs on the sides of volcanoes near the surface of tropical seas —

THE BAHAMA BANKS

The Bahamian Banks pictured here have a profound geological and evolutionary significance. They demonstrate the type of environment in which limestone and its derivatives form enormous sedimentary platforms. They also demonstrate the possibility that early life emerged from the sea and adapted to land in such environments.

Limestone (calcium carbonate) and dolomite (magnesium carbonate) are very common forms of rocks that have been deposited throughout much of Earth's history: for example the pre-Pangean Redwall Limestone, which forms a dominant cliff from end to end of the Grand Canyon, stained red by overlying rocks. In addition there are many other limestone and dolomite formations that make up the canyon's walls.

Most limestones formed in an environment similar to that near the surface of the sea on the Bahama Banks today: for instance those at Cow Head in Newfoundland, about 450 million years old (below, left), or those that cover the sea of salt beneath Mississippi River sediments in the Gulf of Mexico.

Sea level waxes and wanes in response to changes in world climate and Earth's interplate, submarine, tectonic activity. The last Ice Age is a good example of the first. The most active period of mid-Atlantic rifting (when North America was hinging rapidly away from northwest Africa) is an example of the second.

When sea level drops, limestone grains (oolite) and corals are exposed; when sea level rises, corals build fast enough to maintain their critical survival depth below the sea surface. Times of low and intermediate sea levels (as now) are also times for plant life to establish itself on exposed limestone banks.

Mangrove swamps, of which the Florida Everglades seen here are characteristic, exemplify the character of early land-based communities that established themselves on freshly exposed limestone surfaces. Modern mangrove swamps are the habitat of exotic life forms, but they also teem with primitive life forms: freshwater sponges, protozoa, algae, diatoms, sedges, ferns, flatworms, crustaceans — and alligators.

LIMESTONE BANK, COW'S HEAD, NEWFOUNDLAND

MANGROVE SWAMP, EVERGLADES, FLORIDA

in this case the Tethys Seaway. A familiar story of life's struggle to survive was in Darwin's mind. He suggested that, as volcanic activity stopped and tropical volcanic islands began to sink into the seabed, the corals on the volcano flanks had to build their carbonate structures fast enough to stay at the depth in the sea that would ensure the right degree of sunlight. If any part of the coral reef failed to maintain that critical level or if the temperature of the water changed significantly or if the salinity of the water was reduced or if the water became stagnant, that part of the reef would die.

Corals are reef builders of different textures, shapes, sizes, and colors: soft and hard, brain-shaped and branched, fiery red and black — all are tubular carbonate structures produced by polyps (coelenterates, the same phylum as that of jellyfish and sea anemones). Together with the other reef builders, the green and the red fan-shaped algae, the snails, the shellfish, and a host of other forms of life, they produce an average of about 120 tons per square kilometer of almost pure calcium carbonate per year. Carbonates are partly or wholly soluble in water, and strangely they are soluble more in cool than in warm water; their solubility also depends on the degree of carbonate saturation and saltiness of the sea water. Some of the carbonate in coral reefs dissolves, eventually to precipitate as egg-shaped grains of lime, which you and I might call "white sand" but which the geologists call "oolite."

In the Tethys Seaway the faster the seabed sank under the increased weight of what consolidated into limestone, the faster the coral and related life forms had to grow to keep their correct distance from the surface. The increased growth and size of the reefs, combined with the deposition of oolite, and then chemical changes in the limestone beds to form dolomite (magnesium carbonate) developed carbonate platforms that gradually sank into the depths on their volcanic bases at a rate that was accelerated by the crustal thinning caused by the opening of the North Atlantic. This chain of events compounded the carbonate-producing cycle. And the greater in height and area the platforms grew, the more they influenced the flow of surrounding ocean currents. With only occasional interruption all these processes have continued on and around the Bahama Banks for about 140 million years, with the result that the banks have accumulated to a height that core drilling has determined to be in excess of 5,700 meters. The banks themselves, through seafloor spreading, are now far from their place of origin.

San Salvador Island is about 16 by 5 kilometers in extent. We picked up the island on the far southeastern horizon and approached it from the direction of Rum Cay. We passed over an electric-blue patch of sea off Low and High Cays at San Salvador's southeastern end and continued to reduce height and speed the better to take in the view. West of the treacherous coral reefs that littered the windswept eastern shore of the island lay the grayish-white limestone cliffs first sighted by Europeans at 2 A.M. on Friday, October 12, 1492 (this according to an abstract from a manuscript dated 1550 and assumed to be a copy of Columbus's original journal). At 10 P.M. the previous evening the first sighting, by a sailor named Rodcrigo de Triana, was a candle-like light in the darkness ahead. De Triana was on the fast-sailing caravel *Pinto*, which the maritime historian Samuel Eliot Morison estimates to have sped along at about 10 knots in favorable wind; the *Pinto* was ahead of the other two ships, the *Santa Maria* and the *Niña*. At the distance of four hours sailing time the light that de Triana had spotted was most probably the light of a campfire and not that of a candle. At 2 in the morning, after sighting the eastern shoreline cliffs "two leagues from them," the

This diagram illustrates a very early stage in the rifting of Pangea. The Tethys Seaway is shown just beginning to form, and the Atlantic Ocean is an almost completely enclosed sea with a rifting zone down its center. There is a shallow tropical-sea environment at the southwestern end of the basin in which the foundations of the Bahama Banks are being formed above a volcanic region, on the line of a developing transform-fault system.

The drawing of the globe on page viii illustrates the relationship of transform faults to the line of a rift between plate boundaries. Some tectonic scientists believe that this pattern of faulting is related to the centrifugal forces caused by the rotation of Earth about an axis.

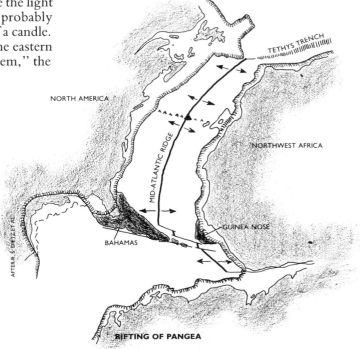

NORTH AMERICA

TETHYS TRENCH

MID-ATLANTIC RIDGE

NORTHWEST AFRICA

GUINEA NOSE

BAHAMAS

AFTER R. S. DIETZ ET AL.

RIFTING OF PANGEA

SAN SALVADOR ISLAND

admiral ordered sail to be reduced and stood off until he could see where best to make a landfall; where that landfall was is quite *unknown*. Most historians favor the western side of San Salvador Island, and from the air a stretch of shore in the lee of the wind seems an obvious choice for so skilled and cautious a seaman as Christopher Columbus had to be.

The beaches of San Salvador Island are an absolute delight: On the lee shore at Columbus Landing there are golden-white sands, quiet, lapping waves, hot sun, cool breezes, and fussy white clouds; but in reality this beach is no nearer to heaven than are ten thousand other lee-shore beaches in the Bahamas. Inland, the airstrip at San Salvador, once pristine and military, is neglected and weed covered, surrounded by impenetrable scrub. The "palm-green shores" on the eastern side of

the island near the cliffs that Columbus sighted were fringed with mahogany trees in his day. But the same cliffs today are flat and geometrically divided, ready to receive what appears to be a huge and ugly housing development.

The cliffs of San Salvador, less than 100 kilometers north of the Tropic of Cancer, seem far removed from the Laurentide Ice Sheet and Hudson Bay. But if it were not for the Pleistocene Epoch and its vicissitudes, today there would be little or nothing to see of the Bahama Islands. Coral reefs flourish best at about 30 meters below the surface, at about 22°C. A few corals grow below both this depth and this temperature, and there are some that survive tidal exposure and bright sunlight. But after allowance has been made for the sinking rate of the gigantic limestone banks (the isostatic rate of depression into the crust below), the present height of dead coral reefs above or below the 30-meter submarine level indicates

the degree of past changes in sea level. At the beginning of the Ice Age, around 2 million years ago, continental ice caps were starting to grow rapidly. At the time of maximum glacial advance global sea level had dropped by 100 meters or more below its present level although, as in other regions, *local* changes in sea level during such periods was influenced by tidal conditions. If today's remnants of the Pleistocene ice sheets melted, sea level would generally increase by 65 meters; and 90 percent of this increase would result from melted Antarctic ice.

Before the ice sheets began to form, the ocean was at its highest level, and the Bahama Banks were covered by sea. Corals formed at their normal depth below the sea surface. Oolite amassed on the lee side of the corals,

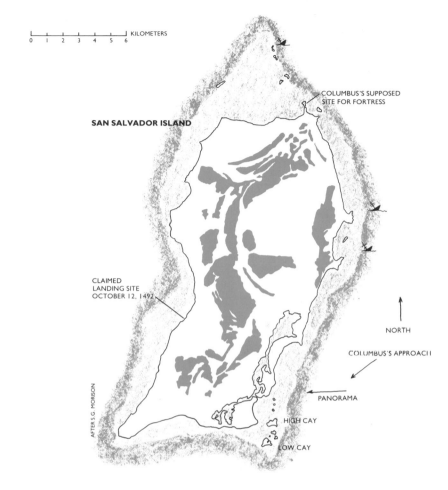

KILOMETERS
0 1 2 3 4 5 6

SAN SALVADOR ISLAND

COLUMBUS'S SUPPOSED
SITE FOR FORTRESS

CLAIMED
LANDING SITE
OCTOBER 12, 1492

NORTH

COLUMBUS'S APPROACH

PANORAMA

HIGH CAY

LOW CAY

AFTER S. G. MORISON

San Salvador Island (formerly Watling's Island), just north of the Tropic of Cancer, is believed to have been the site where Christopher Columbus first landed on October 12, 1492. Columbus sailed on to Cuba and Hispaniola and in later voyages claimed regions of South America for Spain. So far as is known, Columbus did not set foot on the North American mainland.

Exactly where Columbus first landed in the Bahamas is unknown, but if we assume it to have been San Salvador Island, there can be little doubt in anyone's mind that such a skilled seaman would have chosen the lee shore. As can be seen in the aerial panorama (above), the windward shore is strewn with coral reef and toward the right-hand horizon it is also littered with wrecks. The low, gray-white cliffs in the picture are possibly those first seen by Columbus and his men in the early morning light. The most favored landing site is pictured from the air at right and identified on the map of the island.

When Columbus arrived in the Caribbean region, the population included the Caribs, who killed and ate their enemies, thus giving us the word "cannibal" (via the Spanish). The aggressive Caribs of nearby islands and the passive Arawaks of San Salvador were bitter enemies; the Arawaks saw the European visitors as potential allies. Their confidence was naive; for within twenty-five years of their "discovery" most of, if not all, the Arawaks on San Salvador Island and other Bahama islands had been taken by the Spanish into slavery on Hispaniola.

The Bahamas remained depopulated for a century until the arrival of the Eleutheran Adventurers in 1648. The Eleutherans, about seventy in number, were a group of Puritans from England; their effort to colonize did not succeed. Other British settlers from the island of Bermuda later that century had founded Nassau on the island of New Providence, and that town became the capital of the Bahamas.

The ancestors of these West Indian flamingos, which are more brightly colored than are their relatives in other parts of the world, may well have populated the Bahamian Banks since the time of the Tethys Seaway.

David Campbell, a distinguished Bahamian biologist, calls flamingos "desolation's child" because they flourish in regions where few other animals can live — on the salinas, or very salty lakes, that form on some Bahama islands. Flamingos once populated many of the islands, but human beings have decimated them for their plumage and their meat. In the wild they are now mainly restricted to the remote tropical island of Inagua, where they feed on the larvae of the salt-marsh fly and other small organisms. They filter mud and water from the bottom of the salinas through a densely filamented section of their prominent beaks.

"Blue holes" are the now submarine entrances to caves that were worn in soluble limestone by fresh rainwater when the surface of the limestone was well above sea level during the Ice Age (top). Proof that this was so is provided by the recovery of stalagmites from blue holes. Stalagmites are cave formations that can take shape only above sea level, but they have been recovered by divers from depths as great as 45 meters. Dating of one such stalagmite found near Andros Island shows that it was deposited between 160,000 and 139,000 years ago, a dating that corresponds to the time of the Illinoian glacial advance.

The different intensities of blue in the Bahama Banks are caused by depth and by reflected light of mineral content. The photograph shows the colors of the shallow marine environment of a bank of oolite to the upper left beyond the breaking surf on coral reefs. The deep sapphire-blue section shows where the Bahama Banks fall steeply away into great ocean depth and to the abyssal plain of the Atlantic Ocean.

The indigenous people of San Salvador Island whom Columbus met the day he landed called themselves "Lucayans" but are now generally referred to as Arawak Indians. Like other aboriginal Caribbean people the Arawaks migrated from the region of present-day Venezuela in South America during periods of low sea level caused by the last of the great glacial advances, the Wisconsin advance, which reached its maximum about 12,500 years ago. The Lesser and the Greater Antilles as well as the Bahama Islands were larger in area during that time, and the Bahamas in particular are thought to have been idyllic for human occupation. The map reconstructs the Bahama Banks at the Wisconsin sea level compared with present-day sea level.

where it quietly precipitated in the calm shallows (the result of its own accumulation). Tidal bars and sand belts formed by sea currents appeared above surface at low tide. But as the Pleistocene ice sheets grew thicker, sea level was lowered. Coral reefs died, moving their critical life zone down with the changing sea level, while skeletal reefs remained on the surface. Tidal bars and sand belts were exposed and dried by tropical winds and sun. Dry oolite and carbonate debris blown by the wind formed dunes, while former reefs acted as windbreaks that effected the shape of the dunes and the now-extensive islands. Tropical rain percolated through the dunes, dissolving lime and concentrating it at water level. There the particles set like cement, an open-textured kind of "concrete" that thus formed sedimentary rock. The lower the sea level dropped, the greater the thickness of rock that accumulated. Such rock is called "eolianite limestone." It is very soft and easily eroded — so easily eroded that rainwater washed away the more soluble parts of some limestone structures and in some cases formed inland cave systems in which stalagmites and stalactites grew. Some of these have been recovered from now-submarine caves, called "blue holes," on the Bahama Banks. The deepest stalagmites so far recovered from blue holes were found 45 meters below present sea level and are about 150,000 years old.

The Ice Age interglacial intervals waxed and waned, and at times lush forests covered much of Bahamia. People whose ancestors had crossed Beringia thousands of years previously continued the migration by crossing from Amazonia first to the Lesser Antilles, then to the Greater Antilles, and finally to Bahamia. At the end of the Wisconsin glacial advance sea level began to increase. Some aboriginal Bahamians, the Arawaks, remained in what gradually became the separate Bahama Islands in the ever-deepening and ever-widening seaways between the Bahama Banks and the mainlands of South and North America. Some Arawaks lived on San Salvador Island. At some time in the evening of Thursday, October 11, 1492, the San Salvadoran Arawaks lit a fire near Low and High Cays above the eolianite limestone cliffs. On that final pre-Columbian day, with that fire, the Arawaks changed the destiny of their kind forever.

As the shadows grew longer, we flew low to the northeast, just above the continuous coral reefs and rollicking surf on the eastern side of Cat Island. Banking to the west with the wind and having crossed the narrow neck of Eleuthera, we overflew the long stretch of shallow sea above the Great Bahama Bank on our return flight to Nassau. By now I should have expected to be satiated with Bahamian color. But in fact this last leg provided the day's climax. There was a subtle blending of blues and golds of evening sky over the shallow sea, the oolite banks, and the half-revealed sinuous shoals of sand. The color was ethereal. The sensation was one of floating in an infinite dream and was for me one of life's sublime moments.

For all its beauty the Great Bahama Bank's color, like the green of North Cornwall's breakers, the gray off Iceland, and the deep sapphire blue between the Great Bahama Bank and the coast of Florida, is the product of varying mixes of ocean elements: sunlight, salinity, temperature, and water depth. Increased depth, turbidity, and gradual reduction of sunlight produce pastel shades from delicate tints to neutral gray. But seawater temperature contributes to color in an entirely different way: It has life-governing control. The corals and flamingos of the Bahama Islands are at one extreme of colorfullness governed by seawater temperature, and the literally palm-green trees of Oban in Scotland are at the other. Palm trees grow in the open air at Oban, but at precisely the same latitude on Hudson Bay in Canada the ice-cold shores are barren.

A warm, salty stream flows into the Atlantic Ocean between the Great Bahama Bank

CAPE HATTERAS, NORTH CAROLINA

From the perspective of space Cape Hatteras appears to be a tenuous elbow that forms the angle of the bend in the eastern coastline of North America. As can be seen in this panorama taken above the cape looking north, it is in fact far from being an insubstantial thread although it is ephemeral in the geological sense. Cape Hatteras is part of the Outer Banks of shifting sand that form the coastline of what is now North Carolina. The Outer Banks as a whole encloses a portion of a region that was once named "Virginia," after Queen Elizabeth I of England.

The Outer Banks were formed during the Ice Age, when sea level was low and the local winds were no doubt as vicious as they still can be today. Lashed by such winds, sandbanks already exposed permanently above tide level were accumulated into high sand dunes. As sea level rose again at the end of the Wisconsin advance, the dunes became retreating banks of sand that kept above sea level and shifted according to the whims of ocean currents and buffeting waves, which washed their shores at an angle. The Outer Banks and Cape Hatteras have long been a wild shoreline. They became the graveyard of many proud ships.

The gyre of the Gulf Stream from the Gulf of Florida flows along the East Coast shoreline. Off Cape Hatteras the Gulf Stream continues into the Atlantic Ocean, but the Atlantic seaboard turns the Hatteras "elbow." The Gulf Stream was used almost from the time of its discovery by Juan Ponce de León in 1513 to give extra momentum to Spanish galleons and other sailing ships crossing the Atlantic Ocean with the trade winds. Cape Hatteras was their departure point from North America for the Strait of Gibraltar on the same latitude.

and the tip of Florida and distributes its heat around the North Atlantic. It is warmed partly on the shallow Bahama banks but much more on the shelves of the almost circular Gulf of Mexico. The warm Gulf Stream, calculated to be 600 times the volume of the Mississippi River in full flood, is drawn into the cooler, complex mass of the Atlantic Ocean from the Gulf of Mexico. The gulf is a mechanism like the escapement of a gigantic clock, and the Atlantic is the largest of its gyrating cog wheels. The Grand Bahama Bank, fettled in the Tethys Seaway, is the linchpin of the gyres.

The Gulf Stream was first encountered by the Spanish explorer Juan Ponce de León in April, 1513, as he was searching for the "Islands of Beniny" and the aphrodisiac "Fountain of Youth." It seems that Ponce firmly believed in the story of a Bahamian Arethusa's fountain, a fragment of Greek mythology that had somehow been duplicated in West Indian legend: The aboriginals probably invented a rejuvenating freshwater stream on Bimini for Juan Ponce's benefit. According to the official court historian, Antonio de Herrera, Ponce de León sailed northwest from San Salvador Island on Easter Sunday, March 27, 1513, to find Bimini and to quench his thirst. His three ships followed the coastlines of Cat and Eleuthera, missed the turn to the west that would have taken them to the first of their objectives, and sailed beyond Great Abaco Island into the northern extension of the Straits of Florida. Six days later the ships had reached "thirty degrees and eight minutes" latitude (some believe this to have been Ponce de León Inlet, at 29 degrees north, below Daytona Beach) and anchored in shallow coastal water with "a very beautiful view of many and cool woodlands, and it was level and uniform; and because, moreover, they discovered it in the time of the Feast of the Flowers (*Pascua Florida*) Juan Ponce wished to conform in the name to these two reasons."

Having investigated the coast of Florida farther to the north for a week, the three ships went about on Saturday, April 9, and sailed "South a quarter by Southeast," at a safe distance from the coast. It was then that they discovered the Gulf Stream; for the "current was so great it was more powerful than the wind." Although two vessels managed to anchor near the shore, a brigantine was swept out to sea and out of sight. Once reunited, the fleet continued southward, hugging the coast of Florida and so discovering what are now the well-known modern watering places along the Atlantic shore down to the Florida Keys and around the corner into the Gulf of Mexico. But in the years that followed Spanish galleons "coming from the Isthmus," with all the diamonds, emeralds, amethysts, and gold they could plunder, used the Gulf Stream to propel them up the East Coast as far as Cape Hatteras and then well on their way across the Atlantic to Spain at a fast clip to avoid British marauders.

Cape Hatteras is the most easterly point of an elbow of barrier islands on the East Coast. Far below the present sediments of the continental shelf at this point there lie the ancient shores of the Tethys Seaway, and Cape Hatteras is in fact at precisely the same latitude as the Mediterranean Sea. Like the Bahama Islands the barrier islands of Hatteras (the Outer Banks of North Carolina) owe their formation to the time when sea level fell well below its present level. Up to 15,000 years ago a much wider expanse of the passive continental margin shelf was exposed than is now visible. Rivers draining from the Southern Appalachian mountain chain wandered over a wide floodplain to the North Atlantic. As the Ice Age waned and sea level increased, ocean rollers approaching the shallow shelving shore at an angle caused by offshore currents combined with violent Hatteras storms to cause the progressive formation of sandbars that grew sufficiently large to stay permanently above sea level. Sandbars thus became barrier islands, long, ever-shifting, low-profile islands that retreated inland from the advancing sea. On the lee side of the barriers river estuaries continued to flow, and fresh water from the rivers combined with salt water flowing through tidal inlets to form salt marsh and shallow estuarine environments.

Continuing change in the level of the sea caused inundation of the vast area of the coastal plain from the state of Georgia northeast of the Peninsula Arch of Florida to North Carolina. But a second and equally large tract of coastline north of the "elbow" of Cape Hatteras and beyond the Hudson River to Cape

PAMLICO SOUND, NORTH CAROLINA

Behind the Outer Banks of North Carolina are the inner banks and island marshes of a vast estuarine environment. One of the islands behind the barrier in the shallow Pamlico Sound pictured here bears the evocative name of Roanoke, the site of the "lost" Virginia Colony of 1587. All 118 colonists, including two Indian captives returned from England to Virginia, had disappeared without a trace when their governor, John White, delayed by the crisis of the Spanish Armada, returned to Roanoke in 1590.

Access to this apparently tranquil region is through outlets worn by rivers enlarged and made shallow by the action of ocean waves. Approach by sailing ship to inlets in the Outer Banks was particularly hazardous in early colonial days; only small pinnaces were able to make the sometimes dangerous passage.

This is the site of the oldest (1620) British-American settlement yet excavated. It is that of Wolstenholme Towne at Carter's Grove, on the banks of the James River near present-day Colonial Williamsburg. The reconstruction of Wolstenholme Towne's tragic story, no doubt typical of many early attempts to colonize Virginia, is a masterpiece of sophisticated detective work by Ivor Nöel Hume, the resident archeologist of Colonial Williamsburg.

Wolstenholme Towne was a collection of wood-frame and thatched cottages of a design imported from England but modified to make use of locally available materials. The cottages had kitchen gardens and

were grouped near the community barn, the meeting house, and the palisaded fort. The colonists stored their tobacco and lumber ready for shipment to England, grew their own garden produce, and kept chickens inside the fenced areas near their houses.

Hume believes that Wolstenholme was modeled on contemporary Elizabethan settlements in Ireland. But the site of Wolstenholme was in Virginia in North America, *not* Ireland or anywhere else in Britain. The colonists had to learn to adapt to a completely different and to them hostile environment. They were vulnerable to unknown diseases associated with hot, humid, marshy environments, to floods from a large and unpredictable river, and to Indian attacks, against which their settlement proved indefensible.

Cod became even more deeply submerged. The reason for this was that the general tilt of the East Coast from Hatteras to Cape Cod was accentuated by the weight of the Laurentide Ice Sheet upon this section. The ice front crept as far as the Atlantic Ocean at the mouth of the the Hudson River and well out onto the continental shelf farther north. As a result of isostatic depression (from which the region is still rebounding) a series of drowned valleys and embayments was formed when sea level rose, and they are now the natural harbors of Chesapeake Bay, the Delaware River, and the Hudson River.

Under a patent granted by "the Queenes Majestie Elizabeth" on March 25, 1584, to "our trusty and welbeloved servant Walter Raleigh Esquire" Raleigh was given rights "and to his heires and assignes for ever" to anything he discovered in North America, except Newfoundland, which had already been discovered by his half brother Sir Humphrey Gilbert. Raleigh was commanded "herafter to discover, search, finde out, and view such remote, heathen and barbarous lands, countreis, and territories, not actually possessed of any Christian prince, nor inhabited by Christian people...." Raleigh subsequently mounted several expeditions to Roanoke Island in Pamlico Sound, on the lee side of the barrier island just north of Cape Hatteras. An initial exploration resulted in far too glowing a report, but since no other Christian prince had laid claim to the region, Queen Elizabeth felt free to consent that it should be called "Virginia." A mostly military settlement of 107 men was established on Roanoke in 1585. A fort was built, and the surrounding country was thoroughly explored and described. The topogra-

phy of the region and the appearance and customs of its aboriginal inhabitants were extremely well illustrated by John White, a settler who subsequently became governor of the second Virginia Colony. In 1585 local Indian wars were troublesome, and local living conditions were foul. The consensus was that a region 200 kilometers farther north, Chesapeake Bay, would be a great improvement over Pamlico Sound. Apart from fifteen soldiers who were left with provisions for two years to hold Virginia in the Queen's name at Fort Raleigh, everyone else thankfully returned to England with Sir Francis Drake, who had appeared out of the blue on his way to England with twenty-two ships: There was talk of war with Spain.

Another attempt to establish a colony was made by more than a hundred men, women, and children: with John White as governor (and captain) they headed from Plymouth, England, in May, 1587, in three ships, using the privateering route to Chesapeake Bay: the Greater Antilles at Hispaniola, the Bahama Islands, the Gulf Stream, and Cape Hatteras. John White's account speaks scathingly of the ship's master, Simon Fernandes, who he considered to be a careless, ignorant, and deceitful man. White took the smallest of the three ships and sailed through the inlet for a brief visit to Roanoke and the fifteen men who had been left at the fort two years before, but he found only "the bones of one of those fifteen which the Savages had slaine long before." As White, en route to Roanoke, was pulling away from the other two ships, Fernandes "called to the sailers in the pinnesse, charging them not to bring any of the planters backe againe, but leave them in the Island, except the Governour, and two or three such as he approoved saying that the summer [July 25] was farre spent wherefore he would land all the planters in no other place." The colonists had little alternative and never did continue to Chesapeake Bay.

Almost immediately they had problems with the "Savages": Conditions were awful, and their stores were simply inadequate. None of White's assistants would return to England with Fernandes, and they "with one voice requested him to returne himselfe into

CARTER'S GROVE, VIRGINIA

WILLIAMSBURG, VIRGINIA

The British-American Virginia colonists of 1607 were the first permanent British immigrants to North America. They discovered the environment of their settlement at Jamestown, on the banks of the James River, to be disastrous. High humidity and stagnant water bred a super-abundance of mosquitoes. Long, cold, damp winters without adequate housing caused suffering and often death from pneumonia. Inadequate drainage in low-lying marshland caused dysentery and worse. Vulnerability of the settlement to Indian attack completed the colonist's discomfiture and accentuated the differences between the New World and their homeland. A heavy toll was paid in human lives to maintain a riverside community and easy access to supplies from England. It was the better part of a century before one of their number, Francis Nicholson, was able to plan and lay out a new capital that had some of the advantages of home.

The College of William and Mary, chartered in 1693, was the first building to be erected in what was to become the planned town of Williamsburg. The college building was designed in a style similar to that developed by Christopher Wren during his rebuilding of London after the Great Fire of 1666. The basic change in Wren's method was to use bricks instead of the wooden structures that had proved so disastrous in London. The principal buildings in Williamsburg followed this example, but there were grave limitations in the supply of raw materials that led to necessary compromises.

Williamsburg was built on the brow of a hill, above the discomforts of marshland, free from the dangers of flooding, and with obviously improved drainage. The College of William and Mary was built at one end of the hilltop, the State House at the other, and Duke of Gloucester Street between the two. When, much later, the Governor's Palace was built, materials sent from Britain included slate from North Wales already cut to size for roofing. The people of Williamsburg had no more notion of how to slate a roof than the people of Britain had to shingle one. Wooden shingles had by necessity already been substituted for slate.

THE COLLEGE OF WILLIAM AND MARY

DUKE OF GLOUCESTER STREET

MANHATTAN, FROM BROOKLYN HEIGHTS

England, for the better and sooner obtaining of supplies." White had grave reservations but no alternative. Among the 118 people he left on Roanoke were his son-in-law, Ananias Dare, his daughter, Elyoner Dare, and his infant granddaughter, Virginia Dare, "the first Christain borne in Virginia." They were never to meet again: War with Spain threatened invasion by the Spanish Armada, and every English ship worthy of the name was required to be in the English Channel that summer. The Armada was defeated, but John White, despite special pleading to Elizabeth's court, was unable to return to Roanoke Island until 1590. By then the rough wooden houses had gone although the settlement had been enclosed with a palisade. There were no signs of life and no signs of distress — just an enigmatic carving on a bark-stripped tree. Today, apart from the grass-covered outline of Fort Raleigh, there is only a flower-filled Elizabethan memorial garden to remind us of this sorry and incomplete tale.

There is an awful similarity about these early colonial sites: They were marshy, unhealthy, and indefensible. Life expectancy for half the arrivals was forty years, and to reach the age of fifty was an achievement. Because the settlements had mainly wooden structures and were built in estuarine or river-bank environments, their remains are difficult to find. But in the late 16th and early 17th centuries the more inland sites were considered most attractive, as the following description of Chesapeake Bay by Ralph Lane suggests. Here is an excerpt from his report to Raleigh after his return from Roanoke in 1586: "But the Territorie and soyle of the Chesepians (being distant fifteene miles from the shoare) was for pleasantnes of seate, for temperature of Climate, for fertilitie of soyle, and for the commoditie of the Sea, besides multitude of beares (being an excellent good victuall, with great woods of Sassafras, and Wall nut trees) is not to be excelled by any other whatsoever."

Lane omitted from his report that these places were also attractive to local inhabitants and had often been cleared by them for growing maize, squash, and melon. Occupation by colonists growing tobacco, an idea borrowed from Cuba, would lead to bloodshed and massacre on both sides. Meanwhile the colonization of Virginia proceeded but now transferred from Roanoke to the James River in Chesapeake Bay. Although the original site of Jamestown has never been located, its founding established British colonization of the East Coast of North America.

Ivor Noël Hume is as determined, as charming, and as good a detective as one could wish to meet. Hume directed Colonial Williamsburg's archeological-research program and more recently the nearby dig, on the banks of the James River at Martin's Hundred, that revealed details of Wolstenholme Towne (1618), the earliest British-American settlement yet to be unearthed. The story of its discovery, the unraveling of its bloody history, and the extraordinary convolution of clues that led to the archeological solution are

told in fascinating style in Hume's book, *Martin's Hundred*. From Hume and other experts at Williamsburg and later in New York I learned how the modern Pleistocene landscape was used by the colonists and, in particular, how that landscape affected the development of their buildings.

All James River colonists faced major difficulties while trying to recreate their English homes. One was that to transport house fittings across the Atlantic was usually beyond any colonist's means; naturally available materials had to be used. Another difficulty was that Virginia was simply *not* England, it was part of the East Coast of North America and had a radically different environment. A particular difference was that the colonists had settled on a *continental* coastal plain and on river banks of unconsolidated alluvium. This terrain had only passing similarity to the fenlands of Britain, such as those in southeastern England, which are perhaps nearest in character to Chesapeake Bay. Simply everything to do with the environment was on a vastly different scale of discomfort and peril. This was ultimately reflected in Virginian house design:

The city of New York is like London and other major conurbations; it is an amalgam of smaller cities. London is made up from the formerly separate towns of Hampstead and Hackney and Mitcham and a multitude of others now incorporated into Greater London. Many people confuse "London" with the City of London and the West End; these business and shopping centers are a small part of a finely woven web of distribution.

New York City is similarly constituted of smaller parts. The nub of business and shopping in New York City is on Manhattan Island. But unlike London, Manhat-

tan in the 19th century was quite cut off from its neighboring regions by the natural barriers of the very wide Hudson River, the Harlem River, and the East River.

In the 19th century Brooklyn was a major city in its own right on the banks of the East River opposite the Wall Street end of Manhattan Island. The East River could be crossed only by ferry. But New York on Manhattan Island had already been established as the trade center of North America, and ferry boats were just inadequate for Brooklyn's continued development in concert with New York City. A bridge was needed, but it would require a span not previously attempted. After great difficulty and

the surmounting of many technical problems the Brooklyn Bridge, featured in the panorama here, was completed in 1883.

The suspension of the bridge was made possible by the development of spun steel-wire cable. Steel in any shape or form was a novel construction material at that time.

The confinement of Manhattan Island, together with the rapid commercial growth that followed the completion of the Brooklyn Bridge and others, exacerbated the problems of commercial space. The solution was to build upward

instead of sideways. But high buildings were the solution only so long as their floors were easily accessible: Steel cable solved this problem too by making possible the suspension of Otis elevators, elevators having been previously limited to the use of hemp rope and to structures of modest height.

Manhattan is an island of hard metamorphosed rock that was planed by the Laurentide Ice Sheet into a slightly undulating surface. Areas of exposed or near-surface rock alternate with once-marsh-filled depressions in its surface. The configuration of today's skyline in the panorama reflects the original alternation of rock and marsh.

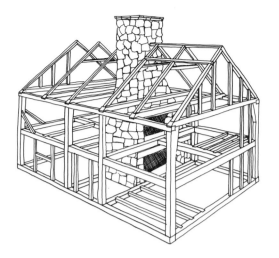

The plans for colonial dwellings were irregular. They were, however, all based on a simple idea imported from Tudor England and adapted to local needs.

A stone central chimney was built first. Frames from wood were constructed separately and raised into position around the chimney. Roof eaves were then made, and the whole structure was pegged together with the chimney as the central support. Thatched roofing, rough plaster-and-lath walls between frames, and windows and doors of the stable-door type completed the assembly. Often these cottages had one room only, with perhaps a sleeping loft above.

This basic design then changed to take the North American envi-ronment into account; the genesis of an Ameri-can architecture had begun.

Great use was made of wood because wood was the most plentiful and readily available mate-rial. At first new designs included additional rooms around the chim-ney, but later design re-verted to the two-roomed, open plan set around an even larger fireplace. This funda-mental approach is still to be seen in many American homes. The reason for the change was that the humidity and heat of summertime is eased indoors when a draft of air is allowed to pass through screened doors and windows. The comparatively small, en-closed, and doored rooms of the later En-glish cottage, built be-tween extra-thick stone walls for insulation, would be insufferable in humid conditions.

One result was an open plan, which allowed any breeze to sweep through open doors and windows — exactly the opposite of "cozy" British cottage construction, which was in-tended to reduce such drafts by constructing segregated rooms with connecting doors. Gradually the colonists began to understand the implication of all the differences.

The imported craft of building a house was based on the concept of a stone chimney to be built first and a surrounding wooden frame structure that was built on the ground and raised into place afterward. A joiner con-structed the mortise-and-tenon-joined tim-bers into braced and pegged frames; he also made the roof trusses and eaves, the parti-tions, the floors (if any), and later the furni-ture. A mason built the chimney; and once the frame house had been erected, he constructed the walls, which were of lime and sand mixed with stone and gravel, built up within the rec-tangular wooden frames. The walls were plas-tered inside and out for durability and normal-ly whitewashed between seasons. Windows were what one could afford: In early Virginia they were either nonexistent or of the stable-door variety; small bottle-glass lattices were a luxury that came much later. But as can be seen in the present, reconstructed Colonial Williamsburg, the basic design of houses was incremental and led to a genetic difference in architectural styles from those of England. For example the original roofs were thatched and liable to catch fire; they were vulnerable to accidental fire as well as deliberate attack. So perhaps one of the most remarkable differ-ences that still distinguish British from Ameri-can house design was the replacement of thatched roofs not with *slate*, cut to size and then fitted and overlapped as in British towns, but with wooden *shingles* used in a similar fashion. When the Governor's Palace was con-structed in Colonial Williamsburg, material delivered from England included a load of Welsh slate cut ready for roofing, but (so I was told) no one knew how to fit them into place.

Town planning was evidently important. In contrast to future developments, Wolsten-holme Towne (excavated by Hume) was based on a style that the English had used suc-cessfully in Elizabeth's time in Ireland: a stockaded fort and lookout, a company com-pound, and a barn, cottages, and kitchen gar-dens — all set near a stream. But the siting next to the James River proved disastrous. The colonists were attacked by clouds of mos-quitoes. They suffered from humidity and heat in summertime and damp, bitter cold in winter. They died from dysentery, malaria, and pneumonia. And they were slaughtered by Indians. The lessons they learned are re-flected in the eventual choice, in 1695, of an elevated position for the site of the College of William and Mary, the first building of what was to be a planned city, Williamsburg, the capital of colonial Virginia from 1699 to 1780. Colonial Williamsburg, now largely restored, is in heavily wooded countryside, with a main street set along the line of a hilltop on the brow of a gentle undulation in the landscape. The James River is nearby, but Williamsburg is well above its flood path and crumbling banks.

New York City, 425 kilometers to the northeast of Williamsburg, is also built by a river. It is situated at the mouth of a drowned valley instead of on a half-drowned fluvial plain. Instead of being composed of crum-bling silt, the banks of the Hudson River were carved and polished into gently undulating slabs by the Laurentide Ice Sheet. One of those slabs of pre-Pangean rock is Manhattan Island. There is no way that those massive and crowded skyscrapers now perched upon it could have been built on the edge of the James River. They would have sunk deep into soft sediments. The skyline of Manhattan dips markedly between Wall Street, at the southern tip of the island, and Midtown, reflecting the once-marshy ground between. But Wolsten-holme Towne and Manhattan have at least one thing in common: Skyscrapers are built just like frame houses, except that the technol-ogy has advanced a little.

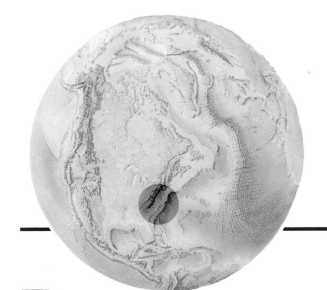

APPALACHIA

RICHARD Hakluyt, the renowned English geographer, published in 1589 a tome entitled *The Principall Navigations, Voiages and Discoveries of the English nation.* In it Hakluyt reprinted "The Relation of David Ingram," in which Ingram brazenly stated that he had walked from the Rio Grande in Mexico to Cape Breton, Nova Scotia — and this in the eleven months from October, 1568, to August, 1569! This was the first claimed traverse of the North American continent by a European.

Ingram stated that, together with a number of others, he had elected to be put ashore rather than remain on board his overcrowded ship. Overcrowding was the result of an un-happy brush with the Spanish off Veracruz in September, 1568, in which the British flag-ship, the *Jesus of Lübeck*, had been lost and oth-er ships in the fleet had become separated. In the months that followed, so Ingram claimed, he and a diminishing number of companions had walked the 4,000 kilometers or more to Cape Breton, supposedly following the North American coastline. Ingram had been picked up and returned to England by a French priva-teer. Ingram's boast aroused the skepticism of distinguished Elizabethans, and eventually led to "Certyne questions to be demaunded of Davy Ingram, sayler, dwellinge at Barkinge in the countye of Essex [and] what he ob-served in his travell."

Every commuter who has ever tried to cross the Hudson River to reach or leave Man-hattan Island during rush hour would prob-ably agree with the words of a 16th-century critic that Ingram's "wretched story is disfi-gured by every form of lie." Even if Ingram had succeeded in crossing the Mississippi and other major rivers during his claimed traverse of the continent, he still would have met an in-superable obstacle in the form of the Palisades and the Hudson River.

One of the problems with building artificial monoliths on an island slab of slightly undu-lating glaciated rock in the middle of a drowned valley is that someone has to invent ways to reach the island. The completion of the Brooklyn Bridge in 1883 solved one such problem: the crossing of the East River from Brooklyn to City Hall at the southern end of Manhattan. No bridge of this size and span had previously been attempted, and the achievement did far more than just make Manhattan more accessible from Brooklyn.

The most significant of the many novel technical decisions made during the planning and construction of the bridge was the decision to use steel cable spun from 10-mile lengths of spliced wire. In the 1850s and 1860s John Roebling had used steel cable in the design of earlier bridges over the Niagara River and the Monongahela River. This was an improvement over wrought-iron links for suspension (such as those that Thomas Telford incorporated into the Menai Strait suspension bridge between North Wales and Anglesey in 1826).

John Roebling died in 1869, and his son Washington assumed direction of bridge construction. As a young man, Washington Roebling had seen canal barges laboriously and dangerously hauled by hemp rope up or down inclines from one stretch of a canal to the next. He proceeded to make stranded steel cable by a method similar to that used for making rope. After experiment he found that steel wire of high carbon content had the flexibility that iron-stranded cable did not have. It is difficult for us to appreciate now that at that time "steel" was a newfangled construction material. The senior Roebling had tried his son's steel cable for suspending bridges of shorter span, and now the younger Roebling determined to use steel cable to suspend the highest, longest, and most exposed suspension bridge so far attempted.

The potential of narrow-gauge flexible steel cable did not escape the notice of Elisha G. Otis, of the Otis Elevator Company of Yonkers, New York. But beforehand, at the Crystal Palace Exhibition in London in 1851, Otis had won acclaim by dramatically demonstrating his "safety hoister," which was raised and lowered by hemp rope. He simply cut the rope while the hoist was half raised, and the hoist automatically locked itself securely in position. Until that time brick buildings had been limited in height to the weight that bearing walls and foundations could support. The construction of the Crystal Palace itself, with its cage-like ironwork of slender beams reinforced by diagonal rods, was soon followed by the building of the first real skyscraper, the ten-story Home Insurance Building in Chicago in 1885: Its metal-frame skeleton released the deadweight from its brick walls. Brick or masonry thus needed to be used only for a skin or paneling, and one of the main constraints on building height had been removed. A few years later Alexandre Gustave Eiffel vividly demonstrated the possibilities by building the 300-meter open-metal Eiffel Tower in Paris. The Otis Elevator Company produced the only quotation for the complete Eiffel Tower installation, and its elevators were suspended from flexible steel cable.

By the end of the 19th century space in which to live and work on lower Manhattan Island was in great demand and short supply; so builders went upward instead of sideways. The use of steel framing had been pioneered in Chicago, but its greatest burgeoning was to be in New York. In the present-day sense of the term lower Manhattan saw its first skyscraper completed in 1908. It was the ornate brick and terra-cotta Singer Building, with forty-one floors and fifteen Otis elevators; at 187 meters it was higher by about 100 meters than was any other building in New York. The building of this skyscraper and its modern counterparts corresponded in method to the constructon of frame houses in Virginia in the 17th century and in Tudor England before that: At the core of the building there was an elevator shaft instead of a chimney; instead of mortise-and-tenon-jointed timbers and wooden pegged frames erected around the chimney, braced and riveted steel girders were constructed around the elevator shaft; instead of an infill of stone and plaster between timber supports, there was an infill, or curtain, of various materials.

Island cities are useless unless they can be reached easily. High structures are useless unless people can get up and down them. Roebling's bridge solved the first problem; Otis's elevator solved the second. The fact that Manhattan was built on a restricted island of peneplained and block-faulted rock has in-

GEORGE WASHINGTON BRIDGE

PALISADES

HUDSON RIVER

THE BRONX

LONG ISLAND SOUND

NEW JERSEY TURNPIKE

PALISADES

MANHATTAN

NEWARK BAY

EAST RIVER

STATUE OF LIBERTY

QUEENS

BROOKLYN BRIDGE

BROOKLYN HEIGHTS

NEW JERSEY

NEW YORK BAY

LINE OF PALISADIAN TAPHROGENY

BROOKLYN

LONG ISLAND

VERRAZANO NARROWS BRIDGE

STATEN ISLAND

ATLANTIC OCEAN

fluenced the building design of cities the world over.

As Joy drove over the George Washington Bridge from Manhattan toward the New Jersey Turnpike and Baltimore, I reviewed what I had learned about the pieces of the geological puzzle that make up the East Coast, the passive margin of the North American continent today. The more I thought about "passivity," the more I began to reflect on what I had previously seen and learned on the West Coast — the active margin and its suspect terrains and mobile belt. I felt that I was beginning to understand what the two margins have in common.

The supercontinent of Pangea, it appeared, was not so much a continent, a single mass, as it was an ever-changing amalgam of continental cratons, pieces of stable platform, ocean crust, island arcs, seamounts, and floodplain lava. Pangea was not an oversized "Australia," but a huge group of islands with shallow seas between its landmasses. There had been no ice sheets for most of Pangea's existence, and sea level was "normal" and therefore much above present level. Sea level had sometimes been even higher than normal, but this depended on the degree of rifting activity beneath the sea. Where continental cratons had collided, the sutures were mountain ranges, and these mountains were all in different stages of erosion at any one time. Where ocean crust was subducting beneath Pangea, there were island arcs and continental batholiths. Huge volumes of subducted ocean crust had caused thinning of the continental crust above them and therefore the formation of basins and block-faulted mountains. Geosynclines and carbonate platforms had formed on Pangean continental margins, and these had been affected by plate movements.

When the bits and pieces moved together to form Pangea and later separated into the present continents, their movement had been chaotic. Individual parts of the whole had rotated and at the same time moved from pillar to post or from pole to equator. If satellite pictures had been taken on motion-picture film at the rate of one frame per *century* for 200 million years during the formation and breakup of Pangea and then were played back at the normal projection speed (twenty-four or twenty-five frames per second), I wondered what the film would portray. It would take

One of the most important geological discoveries of recent times is that the huge tectonic plates that cover the Earth's surface constantly move and interact with each other at plate boundaries. Because of this interaction there have been many tectonic events called "orogenic cycles" during the Earth's long history. These consist of mountain-building episodes and related ocean-basin life cycles.

During the history of Earth there have been orogenies, mountain-forming episodes (part of an orogenic cycle), that can be recognized and approximately dated and geographically located. This can be done because, during deformation and volcanic intrusion, some rocks restart an internal physicochemical "clock" and reset internal "compasses" according to the position of the magnetic pole at the time of their heating. These details can be "read" and interpreted by scientists and thus allow the reconstruction of past geological and geographical events.

The spiral diagram relates specific mountain-building events (orogenies) to specific geological periods, back to Precambrian times. The paleogeography maps show three phases in the subsequent history of Pangea: its full assembly (Triassic Period), its rifting (Jurassic Period), and the time of "rapid" rotation of the present continents away from each other (Cretaceous Period).

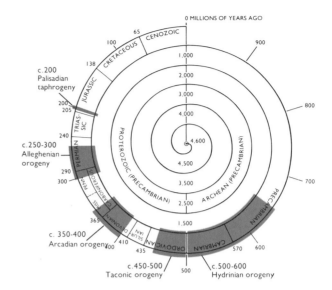

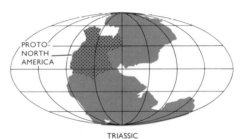

TRIASSIC

JURASSIC

CRETACEOUS

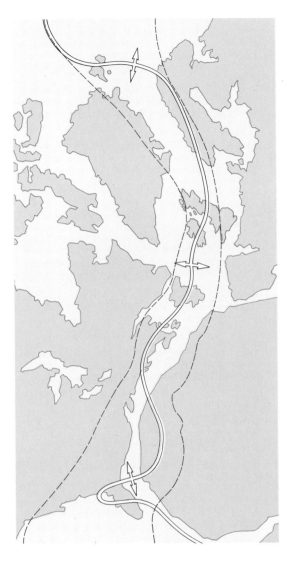

Continental co-relationship diagram, a *schematic* drawing: It illustrates the relationship between today's continental coastlines on either side of the North Atlantic Ocean, which resulted from the formation and separation of Pangea. The true regions of demarcation of continents are the matching outlines of the continental shelves below sea level, and such subsurface features are not shown here.

The line down the center of the illustration represents a generalized boundary between the pre-Pangean continental margins that collided at different times before and during the final assembly of the supercontinent. Some geographic areas to the left and to the right of this line do not have matching rocks because some of them formed on widely separated pre-Pangean continents before the collision.

The broad area between the dotted lines represents a mountain belt, which was a result of continental collisions before and during the assembly.

When Pangea began to hinge apart, its points of separation were different from the original points of suture. This caused "suspect terrain" to be left on the "wrong" side of the Mid-Atlantic Ridge. The red areas on the illustration indicate some of these regions.

In modern geographic terms the diagram suggests that Norway was once a large chunk of Greenland and that it is therefore a part of North America that is now attached to Europe. Similarly, Scotland and Northern Ireland were

also once part of North America. Southern Newfoundland and other parts of the East Coast of North America were formed in the same system as parts of England, Wales, southern Ireland, and northwest Africa. After the division of Pangea parts of what is now North America were left attached to northwest Africa, and parts of what is now northwest Africa form part of present North America.

about twenty-four hours to run the film through, and during that time there would not be a moment when the whole amorphous mass of forming and re-forming continental Pangea remained still. Even if one froze a frame, it would be difficult to choose a moment when the supercontinent was indeed a whole. Seemingly contradictory statements of timing — the Atlantic opening near the British Isles around 60 million years ago and the first rumblings of the "Palisadian rift" in the region of the George Washington Bridge around 200 million years ago — finally began to make sense to me.

The East Coast is now a passive margin separated from its northwest African and Eurasian counterparts by the Mid-Atlantic Ridge, broad expanses of ocean crust, and submerged continental shelves. But I had gleaned from tectonic scientists that in pre-Pangean times, when the North American mass was moving *toward* the African mass, instead of away from it as now and the proto-Atlantic (the Iapetus Ocean) was closing, the East Coast had been an *active* margin. The Iapetus Ocean crust and any suspect terrain upon it had been caught

between two masses about to collide. The Iapetus Ocean bed had been mostly destroyed by subduction in a way similar to that of the post-Pangean destruction of the Farallon Plate beneath the West Coast. But whether the Iapetus floor was destroyed beneath Africa or beneath North America is, I had learned, a matter of debate.

The fragments of terrain caught between "a rock and a hard place" consisted of many pieces, of which two are well known: The first was a structure similar to the Sierra Nevada and Coast Range combination, a batholith and a mélange; and the second was a slate belt, metamorphosed silt similar to the slate belts near the Harlech Dome in North Wales. Together with other rocks the now-leveled surfaces of these terrains form the present rolling Piedmont along which we were traveling to Baltimore. The Sierra-like section follows the full length of the Southern Appalachians from New Jersey to Georgia, whereas the section similar to North Wales forms the slate belt of North Carolina. The exposed geological edge of the Piedmont terrain is abrupt and is called the Fall Line: Wherever rivers flowing from the southern Appalachians run over its edge, there are rock-strewn rapids and waterfalls, which are the limits of navigation for those rivers. The coastal plain of the young post-Pangean formations stretches from the Fall Line to the Atlantic Ocean and continues beneath the sea to the abyssal plains beyond the continental shelf.

The collision of the continents began when the Iapetus Ocean had been reduced to threads between proto-North America and proto-northwest Africa. In satellite film time the act of collision might last an hour or so; in geological terms the process resulted in the Alleghenian Orogeny, a mountain-building period that lasted about 50 million years. The Southern Appalachians formed to the west of the line of suture, and to the east of it the Mauritanides, of which the Atlas Mountains are part. These alpine mountain ranges, or "nappes," were thrown up by the mayhem of the collision. Cratonic rock structures compressed, folded, and overrode each other. The two continents sandwiched bits of Iapetus Ocean floor and its passenger suspect terrains. Figuratively if one compressed the present western section of the North American continent from San Francisco to Denver into a quarter of its present distance, which is equivalent to the distance across the Southern Appalachians from Richmond to Pittsburgh, one would obtain a good reproduction of the geology on the North American side of the suture: And the deformation of the northwest African coastal region to the east of the suture would be almost a mirror image of the same figurative compression.

North America and northwest Africa today are undeniably separated at the Mid-Atlantic Ridge. But where and when the rifting of the two continents began is still anybody's guess. To add to the difficulty, some marginal terrains that originated on one continent are now firmly attached to the other. Moreover, there was no sudden separation but a series of rifts over many millions of years. Nevertheless, some geologists hold the firm view that one of the early murmurings of a break between the continents can *still* be seen by looking up or down the Hudson River from the George

Washington Bridge at the Palisades, the high cliffs that there form the western bank of the Hudson River opposite Manhattan. About 200 million years ago, as in the Basin and Range today, the Hudson Valley region of Pangea was being stretched. The original Pangean "Palisadian rift" is still in evidence, but the block-faulted mountains that resulted as a consequence of the stretching are no longer visible. The Hudson River flows in the Palisadian rift line for a considerable distance: The George Washington Bridge crosses it, and the Statue of Liberty marks its southwesterly progress.

After all that I had seen and learned during my travels I still found some of these geological conclusions mind-boggling. I returned to present reality and gave attention to Joy's driving along the New Jersey Turnpike toward Baltimore. But I couldn't really get away from geology; the New Jersey Turnpike cuts across the continuing line of the Palisadian rift at the Atlantic fringe of the southern Appalachian foothills — a vast swath of land called "Piedmont."

The toll road is part of a multilane highway that adopts other names and numbers en route to Philadelphia on the Delaware River, Baltimore on the Patapsco, Washington on the Potomac, and Richmond on the James — and so to the ill-defined edge of the Peninsula Arch, 1,300 kilometers southwest of New York. These cities mark the limit of upstream river navigation. The coincidence of their geographical location has a geological rationale: They are all positioned on a line of waterfalls and rapids, the Fall Line. This is the intersection where rivers flowing seaward from the Appalachians across the undulating Piedmont meet the flat coastal plain. Early settlements and towns were built at such critical junctions and became the large Eastern seaboard conurbations of today. The smooth, sleek, and swift modern road that links these cities is a wonder of civil engineering, a response to the invention of the internal-combustion engine, the pneumatic tire, and Henry Ford's mass-produced automobile. But modern road building really started with the Romans.

In the first century A.D. the Romans in Britain built a strategic network of straight roads radiating from London. They were so durable that many still exist and often run beneath modern British roads. So from the start of the

Industrial Revolution in the 18th century Britain had at least the makings of a road system. But until John Loudon McAdam developed his paving system in 1815 and exported it to America, roads in the New World were mostly tracks or "corduroy" roads of rough-hewn or -sawn tree trunks or primitive stone roads of little industrial use. Until little more than 150 years ago the best way to move goods and people in North America was not along roads but by water.

Exploring estuaries and rivers was an early colonist preoccupation. To discover a stretch of rapids or a waterfall (and sometimes the colonists discovered a fall-line run in excess of 10 kilometers) was exciting; for fast water meant water power. When the early explorers then discovered the fertile and well-timbered Piedmont land above the Fall Line, they must have been thrilled. The humid, swampy, estuarine and river environments, the James, the Potomac, and the Susquehanna rivers of Chesapeake Bay and the Delaware of neighboring Delaware Bay, served well as ports of entry, but they were hellish places to live in after the moist but temperate Gulf Stream airs of Britain.

Before the development of steam-driven machinery sophisticated use of the waterwheel was the main source of concentrating power into horsepower. Waterwheel power ground grain into flour, forged iron into plowshares, sharpened tools, cut timber, and made roof shingles. The longer the run of the rapid, the greater the supply of water. The greater the difference in level in a river race, the more productive and the more populous the locality became. The largest and the most successful cities developed around the places where the falls were most accessible from deep water — from tidal rivers, from estuaries, and from embayments.

The opening of the West and the threatened completion of the Erie Canal in 1825 made it urgent for the now highly competitive Fall Line cities to have independent waterway links to the Ohio River and so to the Mississippi River west of the Appalachians, a formidable barrier of heavily forested mountain ridges. Richmond was pushing a canal link of the James with the Kanawha and Ohio rivers.

BLUE RIDGE MOUNTAINS AND THE SHENANDOAH VALLEY, VIRGINIA

CYRUS H. McCORMICK'S SHOP, WALNUT GROVE, VIRGINIA

Old World methods proved inadequate to cope with the extraordinary potential of North American wheat-growing areas east and west of Appalachia. The steel plough and the mechanical reaper were the inventions that enabled this potential to be realized.

In the 1830s John Deere, an Illinois blacksmith, developed the first one-piece steel ploughshare, thus making possible the effective ploughing of heavy black prairie soils with minimum horsepower. And in 1847 Cyrus Hall McCormick, a Virginia blacksmith, developed the first mechanical reaper in his shop (pictured left) at Walnut Grove in the Shenandoah Valley.

The steel ploughshare and the mechanical reaper were the two most important 19th-century advances in technology to emerge from the burgeoning New World to the ultimate benefit of the Old World.

The Southern Appalachians are geologically "old" mountains. They have eroded away, and there is little left of their former stature. They are now well rounded and heavily forested, with lush meadows in valleys between folds and overthrusts — such as the Shenandoah Valley and the Blue Ridge Mountains pictured here in late evening light.

The Swiss Alps and the Himalayas are "young" mountains formed by continents in collision (Africa with Europe and India with Asia). Because the present Southern Appalachians were also formed partly as a consequence of a collision of continents, between original North America and northwest Africa during the assembly of Pangea — it is reasonable to suppose that the Southern Appalachians were once Alpine or perhaps Himalayan in scale.

Philadelphia was rooting for the Susquehanna and Ohio river-canal connection. The latter was completed in 1834, but because canals need a head of water to refill their lock chambers, the designers had to overcome the 59 kilometers of watershed between Philadelphia and Pittsburgh. They solved the problem by constructing an extraordinary water-rail transport system, the Allegheny Portage Railroad. Five inclined planes fitted with twin sets of iron rails were built, 413 meters to the summit and 340 meters down the other side to rejoin the canal, which proceeded to Pittsburgh via sixty-eight locks. To transport a barge over the summit ridge, the bargees first floated the barge over a "tram" fitted with wheels set on rails beneath water level at the first stage at the bottom of the incline. The barge was then hauled by rope up to the second stage, while a counterbalancing barge was let down the incline. A steam engine at the top supplied the power. It was this kind of hazardous operation (on the Morris Canal) that inspired Washington Roebling to produce strong, "flexible" steel cable.

While the Philadelphians were planning their link to Pittsburgh, a Quaker milling family of three brothers, John, Joseph, and Andrew Ellicott, had established their prosperous flour-milling town astride the Patapsco River on the Fall Line near Baltimore. To make access to Ellicott easier, they had completed a wagon road to Baltimore. In common with Baltimore businessmen the Ellicotts were worried that the Erie Canal would have an adverse effect on their trade. There was also general concern about Philadelphia's canal plans and the increasing threat from the steamboats that were diverting potential Ohio and Mississippi river trade to New Orleans.

In 1826 a Philadelphia engineer returned from England and reported encouragingly on British railway development. Baltimore's anxiety for its own future increased. Baltimore's fear that the Philadelphians would construct a railroad resulted in 1827 in the incredible decision to build the Baltimore and Ohio Railroad. It was to run west along the Patapsco River valley via Ellicott and so to the Ohio River. This was truly a plan born of des-

THE FALL LINE AT RICHMOND, VIRGINIA

The smooth undulating foothills of the southern Appalachians are called "Piedmont." This very considerable region of North America stretches along the East Coast from the vicinity of the Hudson River to the state of Georgia and beyond. The rocks of the Piedmont also stretch from the Appalachians out beneath the sea to form an underlying section of the continental shelf (see illustration on pages 166 and 167).

The margin between the Piedmont surface and the coastal plain is a very strong feature of the landscape, as shown by this panorama taken at Richmond, Virginia. This feature is called the Fall Line and is literally a line of waterfalls and rapids about 10 kilometers or more in extent at intervals along 1,000 kilometers of coastal plain. The Fall Line marks the limit of seaward erosion of the surface when sea level was much higher than now. The ancient Precambrian and pre-Pangean rocks that form the Piedmont are therefore covered by post-Pangean coastal sediments on the ocean side of the Fall Line.

Many rivers flow from the mountains to the ocean over the Fall Line. As one would expect, the line is therefore the head of navigable waters from the sea. Because of this and because waterfalls provided waterpower to drive waterwheels for industry, cities and towns were founded wherever rivers and Fall Line intersected — notably Philadelphia, Baltimore, Washington, Richmond, Augusta, and many others (map at right).

From the earliest British-American exploration in 1607 of the James River as far as the site of present Richmond until the development of railroads above the Fall Line at Baltimore in the 1830s, waterways were the key to successful use of the prime agricultural land of the Piedmont. Roads were often impassable, and before railroads rivers linked by canals were therefore the only dependable means of industrial development and the only means of connecting the Piedmont with the continental regions beyond the Appalachians: the Ohio River and Mississippi River systems.

Construction of the James River and Kanawha Canal pictured above right was started in 1786. It is an uncanny facsimile of the Leeds-Liverpool Canal, completed in England in 1774; the latter is still in operation, as the red barge shows. The James River and Kanawha Canal, dis-

THE JAMES RIVER AND KANAWHA CANAL, RICHMOND

THE LEEDS-LIVERPOOL CANAL, ENGLAND

used and buried for many years, has been resurrected and restored for the public by local benefactors in Richmond. George Washington was a keen proponent of its construction and in 1784 proclaimed that the James and Kanawha "will be found of equal importance to improve the navigation of both the James and the Potomac . . . [and] of vast commercial and political impor-

tance." During and after completion, however, there were problems of every kind: maintenance and repair of breached banks, deeply frozen surfaces in winter, problems with finance, and from the mid-19th century onward punishing competition from the developing railroads.

THE FALL LINE

ELLICOTT CITY, MARYLAND

B & O SLEEPERS

LOCOMOTIVE NO. I

BRUSSELTON

Ellicott City Railroad Station in Maryland (panorama left) was built near the Fall Line of the Patapsco River at a milling town near Baltimore in the early 1830s. Ellicott was the first terminus of the Baltimore and Ohio Railroad for horse-drawn rail carriages and later for steam locomotives drawing industrial rolling stock. The B & O was the first railroad to be built in North America. Little remains of the original railway line that ran near the modern track seen here. The first track was laid on "stone ties," or "sleepers," of the kind pictured (at the bottom far left) at the B & O Railway Museum in Baltimore.

The first railways in Britain, which inspired railroad development in North America and subsequently throughout the world, were developed in response to the need to haul coal from coal mines. The coal was loaded into "trams" and drawn by horses along a railed track. The hauling was mechanized after the invention of the steam locomotive.

Locomotive No. 1, built by George Stephenson and pictured second from left at the Shilton Works Museum at Darlington, County Durham, England, was used to haul the first successful *public* railroad in the world, which opened at Darlington in October, 1825 — the Stockton and Darlington Railway (1821–1863). The stone block sleepers to be seen in the picture are part of the original course of the Stockton and Darlington Railway. In the next photograph the building at the top of the incline is at Brusselton: It once housed a stationary steam engine that hauled wagons up or let them down inclines on either side of the hill, after which the locomotive took over.

One of the first successful locomotives in America was the B & O's *Tom Thumb*, which had a famous race with a horse-drawn wagon in 1830. The locomotive was supposed to prove its worth against the then-standard means of road transport, the horse, galloping beside the track. *Tom Thumb* lost the race because "the band which drove the pulley which drove the blower, slipped from the drum." But the invention and development of the locomotive and the railroad had too many advantages to allow such minor calamities to stop it.

TOM THUMB

SENECA ROCKS, WEST VIRGINIA

These nearly vertical cliffs, Seneca Rocks in West Virginia, are part of the upturned continental stable platform on the western side of the Southern Appalachian Mountains. The headwaters of the Potomac River (flowing toward the camera) are to the right of the cultivated valley below.

These exposed edges of the continental craton were compressed and folded as a result of the Alleghenian orogeny (300 to 250 million years ago) caused by the collision of the future continents of North America and northwest Africa during the assembly of Pangea. The uppermost of the folds, which once formed an anticline, has been eroded, so that only a lush valley formed from Potomac River sediments, and Seneca Rocks remain.

coach, called the *Experiment*, "commenced travelling on Monday the 10th of October 1825 and will continue to run each day, Sundays excepted." But the primary purpose of the railway was to carry coal, not passengers. With the aid of the engine *Locomotion No. 1*, 10,000 tons of coal were conveyed from Darlington to Stockton in the first three months. The original Stockton and Darlington Railway operated for only thirty-eight years, but its worldwide influence and its influence on the development of North America in particular were incalculable.

The Baltimore and Ohio Railroad Company first experimented with horse-drawn wagons along rails; on one occasion with a single horse it transported two hundred barrels of flour from Ellicott Mills to Baltimore. The company also developed the steam locomotive *Tom Thumb*, one of several competitive steam-locomotive designs of the time. The horses were soon replaced by the locomotive, and the B & O was on its way. A furious war then began between the canal builders and the railroad builders — in fact between the railroad builders and everyone in their way. Such warfare, bullets included, together with engineering difficulties, delayed completion of the B & O route to the Ohio River until 1852. But whatever the human competition or geological difficulty, the railroad age had arrived in America and could not be denied. Industry was now free from the strictures of rivers and the risks of disruption by frozen or breached canals. No more canal systems were attempted through mountains, with or without the help of ingenious rigs. The focus switched to the "iron horse" and the vast weight of iron castings and steel rail required for its development throughout the continent.

Instead of driving directly to Pittsburgh from Baltimore, Joy and I continued southwest on the Fall Line road to the James River at Richmond and there turned northwest so as to cut directly across the Piedmont and then through the valleys and over the ridges of the Southern Appalachians. This choice was in accord with the route of the geological cross section which I had planned the year before and which now appears on page 23. The Blue Ridge next to the Piedmont is of intense interest to geologists because it is part of an overthrust belt in which a section of basement

STRIP MINE, PENNSYLVANIA

peration; for it involved building a track through a mountainous region. The civil- and mechanical-engineering know-how to meet such a challenge had yet to be developed. The backers had no idea of the eventual cost, nor of a detailed route, nor of how they were to cross rivers, nor how they would overcome winter snows and spring floods. They did not even know how they were going to haul wagons along the track.

The decision may have been provoked by the Ellicott brothers, whose milling friends (also Quakers) in Stockton-on-Tees in England had had experience with railroads. The original problem in England had been how to move coal from new mines in the northeast, near Darlington, to the port of Stockton. The canal system was already comprehensive and efficient throughout the industrial center of Britain, but the northeast of England was "too far north" and "too difficult of access" to warrant inclusion in the network. A local decision was made to try steam locomotion and railways instead of a canal and also to encourage the already-demonstrated genius of a young man named George Stephenson. In 1821 Stephenson was commissioned to build a railway from Darlington to Stockton. According to a contemporary poster, a passenger

In pre-Pangean times between 365 and 290 million years ago (see paleogeography and time spiral) great regions of Pennsylvania and other parts of the continent's then-level stable platform were covered with prolific plant growth in a swampy environment. The trees and ferns of this time grew so quickly that new growth covered old before it could rot. As a result peat formed from the vegetation in thick layers, which later were covered by estuarine or shallow-sea sediments. As the sediments grew in thickness, heat generated by the compression of accumulating sedimentary rock turned the peat into coal.

The even layers of many such coal formations were later folded during the Alleghenian orogeny and later leveled again by erosion. A general uplift of the region followed, and this resulted in further erosion of the softer and more vulnerable sedimentary rocks. The result was that coal seams up to 7 meters thick are often found near the surface in Pennsylvania and thus readily lend themselves to large-scale open-cast mining of the kind illustrated here.

Strip mining in Pennsylvania permits 80 to 90 percent recovery of available coal. Draglines with buckets have been developed that can remove 300 to 400 tons per scoop. Some trucks that are loaded from scoops can hold 240 tons and even bigger trucks are now being designed. Today these methods are revolutionizing the mining industry because, after initial costs (including reclamation of ground cover), the cost of strip mining is less than a third that of underground mining. Half the coal mined in this fashion is transported by rail, one-quarter by river barge, and the rest by road.

rocks has overridden a confusion of sedimentary stable-platform rocks, some of which were metamorphosed in the process. We continued northwest up hill and down dale through the alternating heavy forest and lush meadows of the Allegheny Mountains. These were formed from the erosion-resistant, sometimes near-vertical upturned edges of folded stable-platform rocks — the consequence of the geological mayhem that resulted from the collision of North America with northwest Africa.

The Appalachian Plateau between the Allegheny Mountains and the Ohio River suffered the trauma of continental collision but got away with relatively gentle folding and a thorough shaking, in contrast with the violent compressional deformation of the Alleghenies. Subsequently the plateau was eroded to an almost level plain and then uplifted. As its surface elevated, it was eroded into valleys, which often have unstable slopes. The previous peneplaining had reduced the sedimentary rock cover; subsequent uplift and erosion had revealed the edges of coal seams.

I had understood from John D. Haun of the Colorado School of Mines that the bog-water genesis of coal parallels the seawater genesis of oil. In the case of coal organic plant material accumulated as peat in wet, swampy environments in warm, moist, acidic conditions. This contrasts with the accumulation of marine organisms and sediments in briny environments. In the pre-Pangean coal-forming swamps of Pennsylvania plants grew so quickly that fresh growth covered dead material before it decayed. Seven to ten times the thickness of accumulated peat could eventually produce one thickness of compressed and metamorphosed coal. East of the Pittsburgh area thick layers of peat were covered by shallow marine and estuarine sediments. With enough time and pressure from enough sedimentary rock cover the peat turned to coal. The coal was then subjected to further pressure and heat, this time generated by the folding of sedimentary structures during the Allegheny orogeny. The result was that some coal was turned to graphite — almost pure carbon. The hard, high-quality graphite east of Pittsburgh is called "anthracite." Lower-quality coal in Pennsylvania contains more tarry bitumen and is softer. The lowest rank of coal is lignite, and it is nearer to the original peat than it is to graphite. Coal of varying quality underlies more than 10 percent of the North American stable platform. At today's rate of consumption North American reserves will last for three centuries or more.

The city of Pittsburgh was founded in 1788 and was built at the confluence of the Allegheny and Monongahela rivers, which together form the Ohio River at the point of the "Golden Triangle." In the region of Pittsburgh the violent shaking and compression that the stable platform received during continental collision resulted in a landscape that in cross section looks like a jumbled oscillograph line with the top cut off. After uplifting and etching the remaining parts of the landform contained a wide variety of unstable rock strata. Some are horizontal, some are at an angle, some sections are almost perpendicular, and some are bent upward or downward. As a consequence the penalty one can pay for the magnificent view of the spectacle of the Golden Triangle from the heights above the city is the danger of one's house slipping down the cliff into the Monongahela River.

In the moist climate of Pennsylvania the hills, valleys, and cliffs creep insidiously downhill. In arid country, like the Colorado Plateau, landmasses don't creep; they fall or are violently washed away (change there is rare but sudden and complete). In temperate climates, especially in areas of disrupted sedimentary rocks, such as Pittsburgh, erosion causes landslides. These landslides can flow slowly or rapidly, or they can slump or avalanche. They can be caused by talus or soil creep or by earth, mud, and debris flow. Houses built in this region can often be at risk. Geologists, architects, and construction engineers can now largely reduce these risks, but the aboriginal people of the Pittsburgh region did not always recognize the danger.

The dry-as-dust image of an irritable professor is epitomized by Howard Carter, the

One of the most detailed archeological programs to have been attempted in North America was started in 1973 and completed in 1982 at the Meadowcroft Shelter, at Avella, near Pittsburgh. Literally *millions* of bits of information were patiently recovered from a carefully marked and plotted small "shelter" beneath a sandstone overhang. James M.

Adovasio (Chairman of the Department of Anthropology at the University of Pittsburgh) is pictured here opening up the entrance to the now-covered shelter for the author's benefit. Adovasio was leader of the multidisciplinary team responsible for the "dig."

The microscopic to boulder-size data uncovered at Meadowcroft have been collated by computer, have been dated by the latest avail-

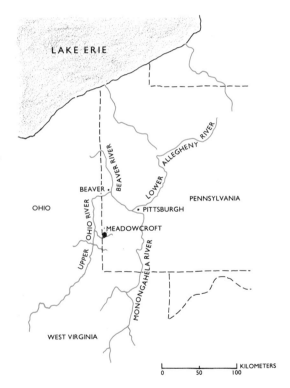

able techniques, and have resulted in a detailed profile of life in this small "overnight" shelter. It was regularly used by aboriginal people from about 19,000 years before the present to about 1,600 years ago. Meadowcroft Shelter is the oldest authenticated site of ancient man in eastern North America.

The map relates the Meadowcroft site to the Ohio River drainage south of Lake Erie; moraine deposits separate the Pittsburgh area from Lake Erie. The cross sections show the development of Cross Creek, in which Meadowcroft is lo-cated, from the time of the Wisconsin advance of the Laurentide Ice Sheet to the present. And the section on the right shows the various sandstone structures that form the Meadowcroft Shelter, with a scale that indicates the modest proportions of the overhang.

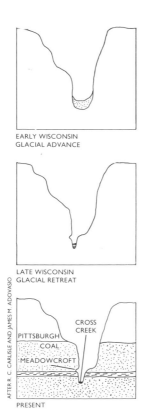

EARLY WISCONSIN
GLACIAL ADVANCE

LATE WISCONSIN
GLACIAL RETREAT

CROSS
CREEK

PITTSBURGH
COAL
MEADOWCROFT

PRESENT

AFTER R. C. CARLISLE AND JAMES M. ADOVASIO

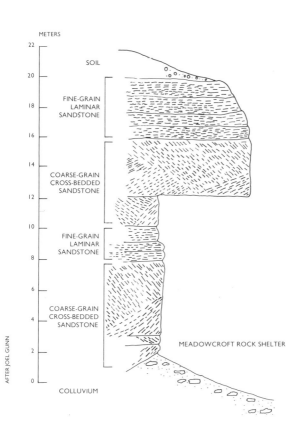

METERS

22

20 — SOIL

18 — FINE-GRAIN
LAMINAR
SANDSTONE

16

14 — COARSE-GRAIN
CROSS-BEDDED
SANDSTONE

12

10 — FINE-GRAIN
LAMINAR
SANDSTONE

8

6 — COARSE-GRAIN
CROSS-BEDDED
SANDSTONE

4

2 — MEADOWCROFT ROCK SHELTER

0 — COLLUVIUM

AFTER JOEL GUNN

MEADOWCROFT SHELTER, AVELLA, PENNSYLVANIA

PITTSBURGH, PENNSYLVANIA

This is Pittsburgh's "Golden Triangle," at the confluence of the Allegheny and the Monongahela rivers, which together form the Ohio River in Pennsylvania. Pittsburgh is one of the most important industrial cities in North America.

Because of its strategic and potential commercial importance the Golden Triangle was the

scene of bitter conflict between British and French forces in the 18th century. Issues were ultimately resolved by the French who burned and abandoned Fort Duquesne, which they had built on the apex of the triangle in 1754. In 1761 the British replaced the French fort with Fort Pitt, named after a famous British statesman of that time, William Pitt the Elder.

Pittsburgh is notable for a number of reasons, many of them having to

do with the geology of the region. Before the city could be commercially developed, bridge design had to advance to span the broad rivers at a number of places — and among these bridges is one of Roebling's steel-cable suspension bridges that preceded the building of Brooklyn Bridge in New York City.

As a consequence of Pittsburgh's extraordinary siting in stable platform folds, which resulted from the collision of America with Africa and subsequent uplift and erosion of the area, many bridges have had to be built to make the Golden Triangle accessible. So many bridges in fact — steel arches, cantilevers, trusses, suspensions, and bridges of unusual design — that Pittsburgh has become a veritable demonstration site for civil engineers and bridge builders.

Local exposure of limestone and iron ore, along with the apparently inexhaustible local supply of trees from which to cut wood for charcoal, encouraged small, independent producers to build smelting furnaces in the early 19th century. Pig iron was produced near river banks upstream, and the product was floated in boats downstream to Pittsburgh for onward shipment. Meanwhile Britain, desperately short of trees for producing charcoal (an essential ingredient of steel-making in the 18th century), had already discovered a way of mass-

producing steel by using materials that also proved to be readily available or easily made in Pittsburgh.

The fact of well-developed steel technology, the simultaneous development of steamboats, the sudden demand for enormous lengths of railroad line and castings for locomotives, and Pittsburgh's comparative ease of access from the East Coast and particularly its ease of access to the whole "western" river system — all these coincidences made Pitts-

burgh the ideal place in which to introduce steel production from "scratch" in North America.

The Bessemer converter of the kind shown at the near right was the key element in 19th-century and early 20th-century steelmaking. The Bessemer converter received molten iron that was converted to steel by a blast of air through vents to purify the metal. Then carbon, manganese, and other ingredients were added in carefully measured quantities. These were reblasted with air until the iron had been converted to steel.

The picture at the far right is of the Ohio River downstream below Pittsburgh: The Ohio drains the western slopes of the Appalachian Mountains and flows into the Mississippi River above New Madrid, a direct distance of 900 kilometers southwest of this point. Upstream of the Ohio's confluence with the Mississippi lie St. Louis and Minneapolis-St. Paul, and downstream are New Orleans and the Gulf of Mexico.

BESSEMER CONVERTER

OHIO RIVER NEAR PITTSBURGH, PENNSYLVANIA

archeologist who opened King Tutankhamen's tomb in 1922. But today that image no longer applies. The modern archeologist is exemplified by James M. Adovasio and his coworkers at the University of Pittsburgh. Adovasio works in an environment of electron microscopes, with computers capable of three-dimensional display, a complex apparatus for determining carbon 14 dates, and a multidisciplinary host of geologists, physiologists, biologists, and other archeologists. But at archeological "digs," as distinct from the laboratory, the fundamental requirements have not changed: There is still a need for patience and for careful, skillful hands. The Meadowcroft Shelter project, started in 1973 and completed in 1982, is a paradigm of such modern archeological research. Adovasio, who led both the field and laboratory research teams, is an athletic, bearded young man who drives a large and commodious truck. He took me to Meadowcroft, where he unlocked the small wooden door now guarding the fully covered Meadowcroft dig. At his invitation I stepped inside and went down a ladder nearly 20,000 years into the past.

There were no gold and lapis lazuli funeral masks at Meadowcroft. But there were hundreds of thousands of minute bits of information compressed into a space no bigger than an ordinary living room. In fact literally several *millions* of individual scraps of information were carefully collected here: sediments and rocks; "tool kits" of projectile points and cutting surfaces; pollen grains, berry seeds, and nutshells; fragments of wood and significant bits of seashell that originated from the Gulf Coast. From the small door at the top, down ladders and wooden planks to the bottom of the dig, I descended back step by step through the years, to the bare rock, where there was no further human trace.

Meadowcroft Shelter is in Cross Creek, near Avella, southwest of Pittsburgh. At the top of the heavily wooded creek Pittsburgh Coal outcrops above a limestone formation. The shelter is beneath even older rock, an overhanging sandstone formation well above the stream that runs to the Ohio River. The overhang has eroded through the ages to form an accumulation of debris and sediments on the shelter floor. The visits of Paleo-Indians who used the site was thus in various ways progressively recorded in the accumulating floor debris. Times of occupation of the shelter have been dated by the charcoal remains of fires and charred animal bone. The oldest recorded date of human presence is about 19,000 years ago.

As the shelter eroded, the roof spalled and distributed rock debris on the floor. As a result of spalling there were at least two major roof falls: the first between 14,000 and 13,000 years ago and the second between 1,650 and 1,350 years ago. The last fall seems to have ended human habitation. Human bone in the shelter is rare: There is one possible burial and a number of bone fragments and teeth. The oldest human bone is a small fragment from a child's hand that has been dated to about 13,000 years ago.

The human use of Meadowcroft was ephemeral and mainly during autumn months. The shelter was seldom occupied for longer than a few days or weeks at a time or by more than a few people at a time. The whole Cross Creek area was heavily used by such groups, and there are many other sites distributed on hilltops in the vicinity. But Meadowcroft's long history is exceptional probably because of its nearness to, and safe height above, the stream running down the adjacent valley. The earliest visitors had tools, which they must have brought with them; for the tools were made from varieties of exotic imported chert. But the visitors quickly began to exploit local Monongahela chert and made single- and double-faceted projectile points and flakes of such chert. These appear throughout the use of the shelter but apparently were not made on the site — which is one of the reasons that led to the conclusion that the shelter was indeed an overnight *shelter* and not a permanent site.

The presence of ash, incinerated bone, modified wood, and basketry suggests that Meadowcroft was used for hunting, collecting, and preparing food. The intensity of use varied, but it was reasonably continuous throughout the 17,000-year period. No Ice Age big-game remains have been found in the shelter, and contrary to expectation the environment of 11,000 years ago (late Wisconsin) proved to be similar to that of today; there were both cool and moist as well as warm and dry periods. But it was considerably warmer between 4,000 and 6,000 years ago. The superficial use of the site did not change appreciably even with the appearance of domesticated plants. Food included nuts, seeds, deer and elk, small game, and bird eggs. Adovasio and his associates were surprised that fishbones did not appear in the sedimentary record until later in the period of occupation, but once they appeared they remained constant until the end, when the roof fell around 1,665 years ago.

Above the limestone formation at the head of Cross Creek there is today a small coal strip mine. I wondered whether coal had been used by the denizens of Meadowcroft Shelter. It seemed a striking coincidence to me that beneath another overhang in the locality of Pittsburg modern man had discovered a similar limestone formation — but this one in association with iron ore. At a place called Sarah Furnace on the Allegheny River drainage, in the 19th century, pig iron was made from the ore by heating it with charcoal and by fluxing it with limestone. Pig iron was shipped downriver to Pittsburgh. Before long the city became the center of massive iron and steel production. By the mid-19th century Pittsburgh was manufacturing rail for railroads, the first steel floors used in Manhattan buildings, and later Roebling's steel cable for the Brooklyn Bridge — which itself started quite a cycle of events.

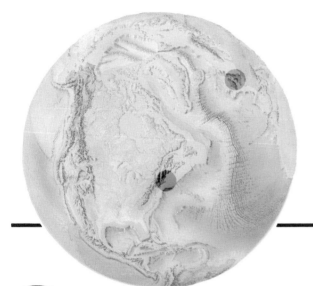

CONFUSION along any line of head-on continental collision is absolute, and its scale Himalayan. Regions on either side of such collisions fracture and ripple. In rippled regions some sedimentary formations are heated to the point of change. As a consequence some siltstone is turned into slate, some limestone is formed into marble, and some coal is converted to "charred" coal. Anthracite is simply compressed charcoal. Because anthracite belts in Appalachia were discovered at a convenient distance from the confluence of two large Ice Age rivers, the city of Pittsburgh became the iron and steel capital of North America. But the first successful iron works built and operated in the New World was on another Ice Age landscape, a modest swamp in Avalonia, at Saugus near Massachusetts Bay.

The *Southern* Appalachian mountains were formed about 300 million years ago, mainly during the assembly of Pangea, in the collision between North America and northwest Africa, a sequence of mountain building known as the Alleghenian orogeny. The *Northern* Appalachians were formed about 440 million years ago, mainly in a sequence of mountain building called the Taconian orogeny, caused by the collision of Laurentia with Baltica, now North America and Northern Europe. But even before these collisions there was yet another period of trauma, the Avalonian orogeny, about 570 million years ago. During this event a large piece of terrain collided with proto-North America; it now forms part of New England. In order to distinguish more easily the geology from current geography, I have here used the expression "Appalachia" to describe the regions associated with the more

recent mountain-forming events and "Avalonia" for the region associated with the oldest event.

Just north of Boston in Avalonia the early British colonists discovered near the head of navigation on the Saugus River a swamp that contained copious quantities of bog iron of the same character as the bog iron the Norse had discovered at L'Anse aux Meadows in Newfoundland over 600 years before. Iron is one of the most abundant elements. During the genesis of Earth most of the planet's iron gravitated to the center and formed part of the core. But huge quantities remained mixed with the floating slag of continental nuclei in still-semiliquid seas of basalt. As Earth cooled and the first sediments began to accumulate in "warm little ponds," elemental iron was precipitated in sediments. But once the gases that formed the primitive atmosphere of Earth contained free oxygen, iron oxide ore began to replace the elemental form. But before this change enormous quantities of iron deposits formed in bands in sedimentary rocks. Some of these formations are more than 3,000 million years old, and they were mostly formed near the surface of continental shields.

The best examples of the older ferrous precipitates in North America are the banded-iron formations of Minnesota on the edge of the Canadian Shield. Much younger banded-iron formations occur in the Pittsburgh region, and early in the 19th century these were the first to be exploited extensively. Pittsburgh also had the advantage of having an adequate supply of limestone and of being near an anthracite belt, although the importance of anthracite was not recognized until the development of steelmaking technology in Britain. Subsequent heavy demand for iron and steel for building railroads in North America and the ease of transporting ore from Minneapolis–St. Paul via the Mississippi and Ohio rivers assured Pittsburgh's future; the city was simply in the right place at the right time.

Saugus Iron Works and its Ironmaster's House were built in 1646 by John Winthrop the Younger, son of the first governor of the Massachusetts Bay Colony. Winthrop manufactured iron there for more than twenty years. The original Saugus buildings have been partly restored by the National Park Service and are now held in perpetuity for Americans as a National Historic Site. Jim C. Gott, the Park Service Superintendent, rounded off my visit with an explanation of the original methods of making iron from bog ore at Saugus.

In appearance the Saugus ore was indistinguishable from the small flakes of bog iron that my fingers had separated from the peat in the peat bog close to the smithy at L'Anse aux Meadows the previous year. At Saugus, however, the ore was so plentiful that it could be

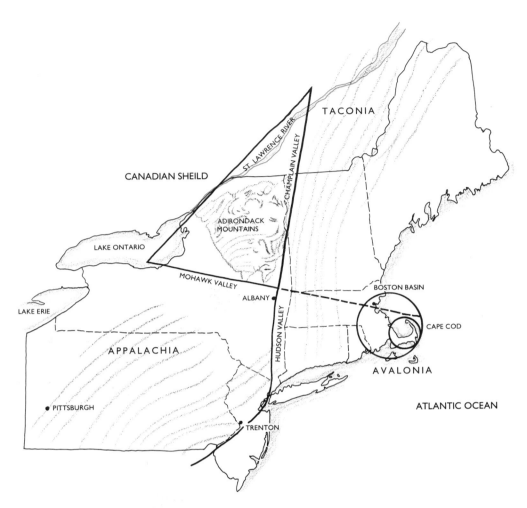

The wedge-shaped Adirondack Mountains inside the right-angled triangle marked on this map are part of the Canadian Shield. The Adirondacks consist of very old rocks of the continent's nucleus that are separated from Canada by a line on the map along the St. Lawrence River and Lake Ontario.

The extended baseline of the right-angled triangle segregates the Northern Appalachian Mountains (upper right) from the Southern Appalachian Mountains (lower left). The Northern Appalachians were formed during the collision of proto-North America with proto-Northern Europe. The Southern Appalachians were formed much later, during the collision of the Euro-American block with northwest Africa during the assembly of Pangea.

The larger of the two circles (lower right) encloses "Avalonia," a region that geologists call the Boston Basin; this consists of rock that existed before the Northern and Southern Appalachian Mountains were formed.

Avalonian rocks were formed mainly in Precambrian and Cambrian times and, as the author describes in this chapter, are of suspect terrain. The smaller circle within the compass of Avalonia contains Cape Cod Bay, which itself was partially formed from Avalonian rock debris during recent glaciation.

This picture is of Saugus Iron Works near Boston, in Massachusetts. The buildings are a modern restoration of the originals, which were constructed by John Winthrop the Younger in 1646. Saugus Iron Works operated until 1684 and so became the first *successful* ironworks in North America.

Because of the difficulty of making steel the Puritans imported most of their requirements from England although they made small quantities of "blister steel" from wrought iron at Saugus. Having shaped this steel in the smithy, cutting edges were hardened by quenching the heated metal in water, a crude form of "case hardening."

The timber for tool handles, for the construction of furniture and buildings, and for making charcoal to smelt iron, came from local forests. But sites like Saugus had a limitation: There would come a time when timber had to be hauled too far. The ironmakers then simply moved to another site near a good supply of wood and iron ore.

The superabundance of natural materials in the New World caused complacency. In England, where forest trees had been used up faster then they could be replaced by natural growth, steelmaking technology was advanced by the necessity of finding a replacement for charcoal. The charcoal substitutes were coked coal and anthracite, and they proved more effective than charcoal for steelmaking because they were capable of producing higher smelting temperatures. Their use led to the development of a process that enabled steel to be mass produced.

By the start of the 19th century the New World was still dependent on the Old for steel supplies and allied technology, but by midcentury the Americans, because of their need to produce vast quantities of railroad lines and steel building structures, made far better use of the Bessemer converter, which had its limitations in Europe.

SAUGUS IRON WORKS NEAR BOSTON, MASSACHUSETTS

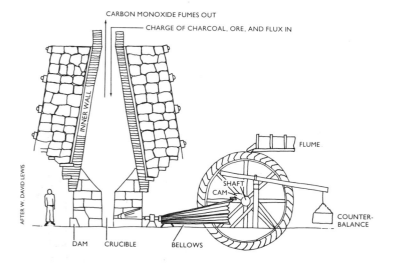

CARBON MONOXIDE FUMES OUT

CHARGE OF CHARCOAL, ORE, AND FLUX IN

INNER WALL

AFTER W. DAVID LEWIS

FLUME

SHAFT
CAM

COUNTER-BALANCE

DAM   CRUCIBLE   BELLOWS

Wooden waterwheels today are a picturesque rural curiosity. But until the invention of the steam engine in the 18th century waterwheels were the main source of power and "were to be furnished in abundance in every parish in the country." This sketch illustrates how the Saugus waterwheel operated the blast-furnace bellows used in smelting liquid iron from a charge of ore, charcoal, and limestone or seashell.

The waterwheel also powered trip-hammers for forged and wrought iron and powered grindstones for sharpening the edges of tools and weapons.

Although the use of steel for tool heads was an advance on iron, just as bronze tool heads were an advance on ancient stone tools, the concept of hand tools had changed hardly at all in thousands of years. This sketch shows how some Stone Age cultures used chert blades secured in wooden handles for axe and hammer heads; they are remarkably similar to modern axes and hammers.

scooped up into flat-bottomed boats from swamp sediments.

Ironmaking is an ancient craft thought to have originated in Asia Minor about 3,000 years ago. Iron was produced by heating lumps of iron ore on an open charcoal hearth and blowing at it with primitive bellows until the charcoal became red-hot and reduced the ore to a soft spongy mass. This was wrought into usable billets by hammering and reheating. Iron ore had defied reduction into a liquid state because it required much greater heat than that required for the reduction of copper and tin ores into bronze, an alloy softer than iron and less useful for making tools. By the time the Massachusetts Bay Company was established in New England and Winthrop built the Saugus Iron Works, waterwheels had been developed to drive oversize bellows by means of levers fitted to the main shaft of the wheel.

Winthrop built a tall chimney-like crucible at Saugus with a replaceable brick lining. Ironmaking was a night-and-day operation, and if the operation were stopped, the inner wall had to be rebuilt. At the start of the proceedings the crucible was charged with iron ore mixed with charcoal and a definite proportion of limestone (or seashell where limestone was not available). The charcoal in the mix was then ignited, and the wheel-operated bellows were used to increase the volume of oxygen available for combustion of the charcoal. The flow of air increased the temperature of the charcoal to white heat — sufficient to "smelt" the iron in the ore, forming a liquid mass that partially filled the crucible.

During the ironmaking process some carbon in the glowing charcoal combined with oxygen in the ore to form carbon monoxide, and this was vented from the chimney above the crucible. Simultaneously the limestone in the mix helped to separate, or "flux," the impurities, which gravitated to the surface of the molten iron in the form of slag. Smelted iron was tapped from the bottom of the crucible and run into a channel with side branches formed in sand on the ground. The main channel was called a "sow," and the small branches off it were called "pigs." Such cast pig iron was brittle when cold and usable only for making items such as pots, pans, and stoves.

FLANNIGAN'S ISLAND, NEW HAMPSHIRE

To produce iron suitable for making tools and weapons that were to be later fitted with handles of wood, pig-iron billets were wrought into malleable iron by a blacksmith, who worked a small proportion of slag back into a reheated billet by folding and hammering and reheating and refolding. This was done with a mechanical hammer driven by wooden reciprocating levers and gears powered by the waterwheel. An alternative to wrought iron was steel, a harder, stronger, and more flexible version of iron and therefore desirable for sharper-edged tools and weapons. But to produce steel, it was necessary to incorporate a great deal of carbon with the molten iron during smelting, and this required even higher temperatures and more charcoal in the crucible. In colonial New England steel was made in small quantities by reheating strips of wrought iron covered with a sprinkling of charcoal. The strips were baked in a sealed clay vessel to form what was called "blister steel." The process was both expensive and exacting and it was thought preferable to import steel from Britain.

In New England and elsewhere in North America the raw materials for producing cast and wrought iron — waterpower for waterwheels, trees for charcoal, limestone or seashell for flux, and iron ore for smelting — were readily available and close to each other at convenient sites. The overabundance of raw materials put the industrialists of the developing continent into a false position of security. In Britain, however, even by the early 17th century, the concentration of the developing iron and steel industry had denuded the comparatively limited forests of the land faster than they could be regrown. The forests had also been thinned by the need to build ships in great numbers: for instance to build the fleet that had defeated the Spanish Armada in 1588.

In England it became necessary to experiment with coal as a substitute for wood for heating homes and for smelting iron. But it was found that bituminous coal contained such impurities that it could not be used to replace charcoal either for smelting iron or for producing the higher crucible temperatures necessary for large-scale steel manufacture. Shortage of wood to make charcoal and yet an abundant supply of coal underfoot provoked continued effort to solve the problem. But the solution was not found until the development of *debituminized* coal, "coked" coal, in 1709.

There is a reason for the vivid autumn foliage shown here at Flannigan's Island in New Hampshire and common throughout this and neighboring states. The color is the result of extremely cold winters, moist springs and summers, and hot, dry, still autumn days with subfreezing nighttime temperatures.

GUARD LOCKS, LOWELL, MASSACHUSETTS

This picture is of the Guard Locks above the Pawtucket Canal, which supplied water and waterpower to North America's first planned industrial community at Lowell, in Massachusetts, about 30 kilometers northwest of Boston. Quite remarkably the *whole* of this industrial city is now being restored by the National Park Service as a National Historic Park.

Lowell was founded in 1821. The reason for its founding is explained by the 1832 map reproduced above. The wide Merrimack River makes a 90° turn around the locality of Lowell. Just below the place where the Pawtucket Canal joins the Merrimack River, the formidable stream flows over Pawtucket Falls. Here the river level drops through about 10 meters in a few kilometers. It was dammed so as to redirect water from the river into the Pawtucket Canal and to provide canal-boat passage up and down the local fall line.

The founders of Lowell bought out the proprietors of the canals and the land adjacent to the river, thus gaining control of the waterpower at the falls. They formed the Merrimack Manufacturing Company and transferred the property title to the company in 1822. They then enlarged the canal to convey water to cotton mills that they built, equipped, and had operating by 1823. Within 40 years 380,000 spindles were producing cotton. *All* the machinery was driven by waterpower over Pawtucket Falls. Here the river level drops through about 10 meters in a few kilometers. It was dammed so as to redirect water from the river into the Pawtucket Canal and to provide canal-boat passage up and down the local fall line. and was operated by British and later by French, Polish, Portuguese, Greek, and Jewish immigrants.

The Lowell cotton mills were serviced by an 8-kilometer network of canals. In time the supply of waterpower was exhausted and the large mills had to adopt auxiliary steam power to maintain commercial growth. This was a reluctant step because steam power cost three times as much as waterpower. But the step was inevitable; it was in accord with North American diffusion of the 18th- and 19th-century Industrial Revolution in Britain.

MERRIMAC RIVER

MANUFACTURING

CANAL

CANALS

BOARDING HOUSES

MANUFACTURING

PAWTUCKET BRIDGE

MASSACK FALLS

CANALS

FALLS AND WEIR

PAWTUCKET CANAL

BOARDING HOUSES

CONCORD RIVER

RAILROAD

GUARD LOCKS

LOWELL, MASSACHUSETTS

The use of coke in place of charcoal and the later discovery of the importance of anthracite advanced iron and steel technology in Britain to an unassailable position. The discovery of coking and anthracite to replace charcoal was almost simultaneous with the discovery of steam power to replace waterpower. These discoveries set in motion the 18th-century Industrial Revolution in Britain.

In contrast with the dirt and grime of industrial England in later years Elizabethan England was still an agricultural society and was also a remarkably enlightened place in which to live. Attitudes in England had been molded by Queen Elizabeth herself. She traveled widely and met her people in their towns and villages; she was a much-loved and tolerant ruler, an exception in despotic and bigoted, divisive times. Her subjects had learned to steer a reasonable middle course in voicing both their political and their religious opinions. Extreme views were harshly discouraged, but moderate views were unfettered. Disloyalty or intrigue was punished severely, sometimes by death. But the strand of intellectual and theological moderation was still exceptionally broad for its day.

Elizabeth came to the throne after a previous reign of harsh intolerance. To reunite her kingdom, she had carefully cultivated an ill-defined flexibility of state and church government. But after her death in 1603 her tolerant policies were undermined by increasing bigotry, which again polarized the country.

Elizabeth's second cousin, King James VI of Scotland, who succeeded Elizabeth and became James I of England, had been tutored in rigid Calvinist doctrine; but as he later loudly announced, he had found this tutelage insufferable. Although much admired for his theological scholarship in Scotland, his political education by Scottish lairds in a close community had not prepared him for the rigorous hierarchy of the English court. His Yorkshire father, Lord Darnley, had been murdered; his Catholic mother, Mary, Queen of Scots, whom James had not seen since infancy, was unwelcome in Scotland and had been confined in England by her cousin, Queen Elizabeth, for twenty years. Mary appeared to invite her own eventual execution by persistently plotting to gain the English throne through the murder of her cousin. Against this dark family background James's parliamentary ideas had been shaped by his experience with a Scottish assembly that endorsed absolute monarchical views: It did not discuss and perhaps reject them as Parliament had done in England.

During his progress south from Scotland, through the North of England down to London, for his enthronement in 1603, Puritans who conformed to the English established church petitioned their future king. Their "Millenary Petition" was said to have been signed by a thousand English churchmen who wanted to have their ecclesiastical position legally confirmed under church law rather than to continue under ambiguous Elizabethan "toleration" into the new reign. James saw in the petition an opportunity to advance both his reputation as peacemaker and his popularity in England, an England that was generally distrustful of foreigners.

Soon after his coronation James took steps to end England's war with Catholic Spain. He must have started discussing the prospect of American colonies, for this was a pressing subject: In 1602 and 1603 several further but abortive attempts had been made to return to Roanoke Island near Cape Hatteras, with depressing results. Then James decreed a Hampton Court Conference, to be held in January, 1604, under his presidency. The convocation was to examine the "Millenary Petition" in the presence of bishops and representative Puritans. James did not follow the example of his predecessor in dealing with confrontations; Elizabeth had managed them with a deft combination of power and procrastination, of indecision with wit.

Early one cold, misty, frost-rimed morning I stood shivering in the silent solitude of a cobblestoned courtyard. It was here, at Hampton Court Palace, on such a morning nearly four centuries before that two communities in New England in the New World were smelted and wrought. I imagined the scene as the convocation gathered for its final meeting: I could almost hear the clopping of horses' hooves hard on the cobbles, the clink and clatter of breastplates, the curt military orders, the closing of doors, the footsteps fading to faint echoes down stone-flagged corridors and the silence that followed them. Then the muffled murmur of men, a shout maybe, and voices raised in a sudden clash of opinion. The babble of approach. Doors flung open. Angry, flushed faces as Puritans departed. Smug, smooth bishops now secure in their power.

James at first adopted a friendly disposition at the conference. He had rebuked Bishop Bancroft for trying under canon law to deny the Puritans' right to speak. But in the face of increasing Puritan rudeness, particularly from the Scottish members present, he had decided point after point against them. The final proposition made by the Puritans had been that the lower clergy should have the right to meet in conference and that the bishop should consult the synod of his diocese. This apparently was the spark that caused the explosion. The word "synod" reminded James of his daily humiliations as a youth. The King in a sudden rage dismissed the conference with the words "If you aim at a Scottish Presbytery, it agreeth as well with a monarchy, as God and the Devil. How they used the poor lady, my mother, is not unknown, and how they dealt with me in my minority. I thus apply it.... No Bishop, no King.... If this be all your party has to say, I will make them conform themselves, or else will harry them out of the land." These angry words were a turning point in British and subsequent North American history.

The outcome of the Hampton Court Conference must have been eagerly awaited by many who did not attend. And a few of these had views that no one but themselves believed would be even broadly acceptable to the King. These were the Separatists: They wanted nothing to do with compromise and conformity within *any* established church. They wanted to practice their theological beliefs without constraint. They rejected all convention: the marriage ceremony and the exchange of rings, the sign of the cross, genuflection,

HAMPTON COURT PALACE NEAR LONDON, ENGLAND

Hampton Court Palace near London, England, is here shown on a sunny day in spring. The palace is redolent with memories. Its ancient interior fabrics, its Holbein portraits, its extensive herbaceous borders, its wisteria-covered walls, its vinery, the smell of fresh-mown lawns, and even the slightly musty aroma of the River Thames itself — all these cannot have been so very different four centuries ago.

But it was on a cold, dank, stark January morning in 1604, within these walls, near this river, that the future history of North America was to be given an important twist. The leading role was played by King James I of England (who was also James VI of Scotland), presiding at the Hampton Court Conference. James had called the convocation to debate the "Millenary Petition" with bishops and Puritans, and the debate deteriorated into an angry confrontation.

One of the contemporary accounts of the event was written and published in 1640 by Izaak Walton, better known not as an historian but for his masterpiece, *The Compleat Angler*. Walton wrote of the conference that "[t]here was one *Andrew Melvin*, a Minister of the Scotch Church, and Rector of St. *Andrews*; who, by a long and constant Converse, with a discontented part of that Clergy which oppos'd

Episcopacy, became at last to be a chief leader of that Faction: and, had proudly appear'd to be so, to King *James*, when he was but King of tha Nation [that is, of Scotland], who the second year after his Coronation in *England* conven'd a part of the *Bishops* and other Learned Divines of his Church, to attend him at *Hampton-Court*, in order to a friendly Conference with some DIssenting Brethren, both of this, and the Church of *Scotland*: of which Scotch party, *Andrew Melvin* was one. . . .

at *Hampton-Court-Conference*, he there appear'd to be a man of an unruly wit, of a strange confidence, of so furious a Zeal and of so ungovern'd passions that his insolence to the King, and others at this conference, lost him both his Rectorship of St. *Andrews*, and his liberty too: for, his former Verses, and his present reproaches there used against the Church and State, caus'd him to be

committed prisoner to the Tower of *London*: where he remained very angry for three years."

As a result of Melville's (Walton calls him "Melvin") indiscretion and rudeness and that of others who attended the conference, King James threatened to "harry out of the land" any clergy who did not conform with his Episcopal Church.

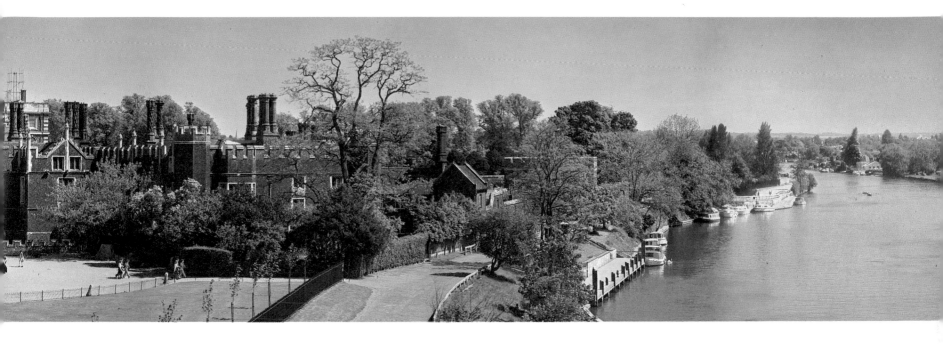

the celebration of Christmas Day, sport on the Sabbath, the use of "amen" to end prayers, and many more. The Separatists were considered rebellious and indiscreet even by the Puritans. They became dangerous to know. But they learned to be cautious; they met secretly at a manor house near a Roman road in Scrooby in Lincolnshire. Some of them later became the Pilgrims of the "Plimoth Plantation" on the shores of Cape Cod Bay in New England. They were later followed to New England by the more restrained Puritan members of the Massachusetts Bay Company, who settled Boston and started an iron works at Saugus.

From London Joy and I went by the modern road that runs over the Roman road, starting from the City of London at Bishopsgate, across the fens to Lincoln, northwest of Boston on the shores of The Wash. The road continues to York and to Hadrian's Wall and beyond — a road that travels almost to Edinburgh and the Firth of Forth in Scotland. This Roman road became a post road for royal

messengers, and in the 16th and 17th centuries it was used as the main monarchical "progress" route and is for this reason called Ermine Street. Other roads were subsequently constructed on top of the Roman network of roads or near them: macadamized roads, concrete motorways, and steel railways from London to Scotland. And all because the eastern network was constructed on part coastal plain and part piedmont between the "backbone of England," the Pennines to the west, and a shelving, shallow, sea-filled graben — the epeiric North Sea.

My immediate interest was a left-hand turn just beyond Lincoln that leads to Scrooby. This now little-known country road is a stretch of an earlier Roman road that led to an easy crossing of the Don and Humber rivers and then to Leeds in Yorkshire and to York Minster, the seat of the Archbishop of York's diocese. Like literally millions of other British commuters to London from the north, by the Great North Road before the present motorways were constructed or by the mainline railway that passes within a few hundred meters, I had been within sight of Scrooby innumerable times without knowing that it existed. But at this small community of Scrooby in January, 1604, the bad news from Hampton Court hurt most.

Malcolm J. Dolby, curator of archeology and history at Doncaster Museum, lives with his young family in a tiny Elizabethan cottage in the present country village of Scrooby, within a few hundred meters of the Scrooby Manor House site. Dolby told me some of the results of his research into the history of the locality.

The use of the Scrooby Manor House site for matters ecclesiastical dates back quite a while. The record shows that in 1207 King John ordered wine to be sent to Scrooby Manor for his half brother Geoffrey, Archbishop of York, and that later King John had himself visited the place. Successive archbishops had a huge diocese to govern, and Scrooby became one of the centers from which episcopal tours of the diocese were conducted. The earliest brick building at Scrooby was constructed at the turn of the 15th century, and part of that fabric was incorporated into the present remnant building. Visitors after completion of the manor's original building included Henry VIII's disgraced secretary of state, Cardinal Wolsey, who had built Hampton Court Palace in 1520 and had given it to his king (al-

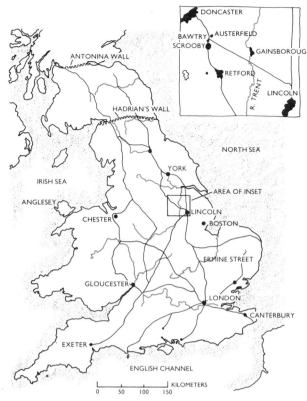

ROMAN ROADS OF BRITAIN

The Anglo-Saxon name for one of the main Roman roads in Britain is Ermine Street because one Roman road, now mainly covered by modern roads, was used by the royal households of England during their "progress" from London to the north of England. This simplified map of Britain shows part of the substantial network of Roman roads built during the Roman occupation of Britain. Many of these roads still exist nearly 2,000 years after they were built.

In the 17th century Ermine Street was a post road for royal messengers. Scrooby Manor House in Lincolnshire was a post office on a subsidiary Roman road from Lincoln to York, the seat of the Archbishop of York. The bad news of King James's threat to harry unconformists out of the land reached Scrooby Manor House via Ermine Street. The postmaster at Scrooby in 1604 was William Brewster, a future Pilgrim Father.

though this had not prevented his later execution). Long after Wolsey, in 1541, Henry had himself visited Scrooby Manor and had held a Privy Council there. Scrooby must by then have been a magnificent moated building in the finest Tudor style; but it was destined to be mostly destroyed in 1636 by order of Charles I.

After an overnight stop at an old coaching inn in Bawtry we returned to Scrooby village, intending to photograph what is left of the former manor house. I had learned from Malcolm Dolby that one wing had been renovated two centuries ago by the Archbishop of York for the use of a tenant farmer. I wanted to capture the spirit of this remnant in the low light and mist of an early morning, but that "spirit" proved to be beyond redemption. What little remains of the original manor is no more than a souvenir of King Charles's petulance. The former grace and style of a Tudor manor house is now reduced to a sad commemorative plaque on an unkempt farmhouse building with a companion galvanized-metal cowshed. But beneath the cowshed and the folds of the surrounding fields there must be a rich trove of history for someone to unearth.

I wandered around the ramshackle buildings, looking for the old course of the River Ryton, which was part of the original manorhouse moat. I ducked beneath a barbed-wire fence and fought a forest of reeds and there it was, filled with weeds, an odd piece of waterlogged timber floating beneath the surface. Before the dark days that followed the Hampton Court Conference the manor house was occupied by the Archbishop of York's bailiff, William Brewster. His father's episcopal appointment and that of master of Queen Elizabeth's post had been awarded to him after his father's death in 1590. The year before, Brewster had returned to Scrooby from the Palace of Westminster, where he had served as personal secretary to one of the queen's diplomats with responsibilities in the Netherlands; but it seems that Brewster was bitterly disillusioned by the way of life at Elizabeth's court. About the time of his return to Scrooby a young orphan from the nearby village of Austerfield, William Bradford, was invited to live at the manor: Together, Brewster and Bradford were destined to become the chief architects of the Plymouth Plantation.

After the Hampton Court Conference James I appointed Bishop Bancroft Archbishop of Canterbury and encouraged the strengthening of the laws against religious nonconformity. This aroused great resentment and defiance. Parliament was not asked to ratify the new laws and therefore championed those clergymen who lost their livings as a result of them. William Bradford and others who took extreme views, such as John Robinson of Sturton-le-Chapel, became members of a Separatist church formed in Gainsborough. And when John Smyth, the minister of that church, was "harried out of the land" in 1606 and fled to Holland, Scrooby Manor became the secret meeting place for his congregation, with John Robinson as its minister. This congregation too was sought and harried, and as a consequence most Scrooby Separatists in turmoil left with their families to join Smyth in Holland. In December, 1607, Brewster and the few remaining members of the congregation at Scrooby were sought by the Ecclesiastical Court for nonappearance. A court officer sent to arrest them "certifieth that he cannot finde them nor understand where they are." Thirteen years later, years of uncertainty and growing resolve to start afresh, Brewster and fellow British expatriates arrived at Cape Cod in the New World.

Cape Cod is an extraordinary shape and is a relatively new feature of the coastal landscape on the New England continental shelf. The cape was formed as a result of rocks and till moved by Laurentide ice. Martha's Vineyard and Nantucket Island, due south of Cape Cod, are terminal-moraine features of the Wisconsin advance, whereas Cape Cod itself is the result of an interval of colder climate during the general retreat of the ice sheet. As I had learned in Wisconsin, ice sheets do not retreat in the active sense; they simply stop advancing and begin to melt without further forward movement. The large accumulation of rock at the ice front is deposited as moraine;

Perhaps one of the least known but important historical sites in England is illustrated here. The picture shows all that remains today of Scrooby Manor House in Lincolnshire — now an ill-kempt farmhouse between a main railway line and a motorway. But Scrooby Manor once resembled the sketch of another medieval moated manor house still in existence, Otley Hall in Suffolk.

At the time of the Hampton Court Conference in 1604 Scrooby Manor House was occupied by William Brewster and his family and by Brewster's young protégé, William Bradford. With others they formed the congregation of a Separatist church at nearby Gainsborough (see map). Separatists were *not* considered "Puritans" in their time. Separatists were not prepared to compromise in *any* way with *any* estab-lished church; the Puritans considered Separatists to be extremists and too dangerous to know. The Separatist church at Gainsborough was closed by the Ecclesiastical Court, and its minister, John Smyth, fled to Holland. But his congregation continued to meet secretly at Scrooby Manor under the ministration of John Robinson and with the encouragement of William Brewster.

At that time Brewster was not just postmaster for royal messages on their way to York but also bailiff of the Archbishop of York's diocese in the region of Scrooby: The manor house belonged to the archbishop. Many of Brewster's brethren were also scholars of Cambridge University in the nearby fenlands. Their Separatist views and their use of Scrooby Manor for meetings were both defiant and dangerous, to put it mildly: They were indeed *certain* to be harried from the land.

SCROOBY MANOR HOUSE, LINCOLNSHIRE, ENGLAND

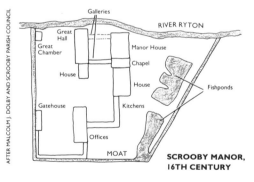

AFTER MALCOLM J. DOLBY AND SCROOBY PARISH COUNCIL

Galleries

RIVER RYTON

Great Hall

Great Chamber

Manor House

Chapel

House

House

Fishponds

Gatehouse

Kitchens

Offices

MOAT

SCROOBY MANOR, 16TH CENTURY

OTLEY HALL

CAPE COD, MASSACHUSETTS

stagnant ice forms kettle lakes of various sizes and deposits erratics. Braided streams of meltwater deposit an outwash plain beyond the moraine.

On Cape Cod, Martha's Vineyard, and Nantucket the moraines follow the curvature of the original ice-lobe fronts, and therefore the highest ground is formed on the cape and its related islands. The outwash plains form the flat, apron-like, lake-riddled landscape that extends seaward. Some erratic boulders found on Cape Cod are composed of pink granite and have been matched with rocks 250 kilometers to the northwest. Another, better-known erratic, Plymouth Rock on the western shoreline of Cape Cod Bay, has had a checkered history. It is composed of grayish

granite and was broken and repaired several times when it was moved for use as a symbol of liberty. The Laurentide ice may have carried it to its postglacial positon from as far away as New Hampshire.

After the end of the Wisconsin advance the rough, rocky, and pitted postglacial landscape of the continental shelf in the region of Cape Cod was slowly inundated by the rising level of the sea. Sea erosion then caused sandy bays to form ragged scalloped shorelines. Continual back-and-forth dragging action by waves in shallow water caused sand to move laterally along the shore for considerable distances to form spits, of which Cape Cod is the perfect example. The quiet bays inside the old moraines became shallow and sheltered while the active beaches on the ocean side of the outwash plains were made dangerous for sail-

ing vessels by the shifting shoals and sandbars. Rain falling on the sandy soil above sea level formed a freshwater lens that extended approximately forty times deeper beneath the ground than it stood above it. Some kettle lakes dried up, and cranberry bogs formed in their place. The bogs accumulated peat, which much later was used as fuel. The soils the colonists found on Cape Cod and its vicinity were similar to those found in other heavily glaciated northern regions and were a poor substitute for the rich loam of Lincolnshire.

Cape Cod proved untenable. The beckoning finger of the cape could not be approached by ship from the Atlantic side. The shelter the

Cape Cod's landscape was formed from glacial moraine eroded by waves and wild winds. The Atlantic Ocean pounds the outside shore of the cape, and the sandy sediment in its shelter forms shallows in Cape Cod Bay. As the two diagrams show, the Laurentide Ice Sheet advanced southeastward carrying rock debris carved from the mainland onto the then-exposed continental shelf. The ice sheet stalled and melted. Rocks, glacial till, and huge blocks of rotting ice were deposited to form Nantucket Island and Martha's Vineyard. There were a temporary advance of the ice front and then a final melting that left the profile of Cape Cod much as it now appears — but since whipped by 10,000 years of roaring Atlantic surf. As the Pilgrim Fathers discovered in November, 1620, this was no place to found a settlement.

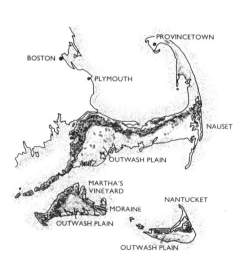

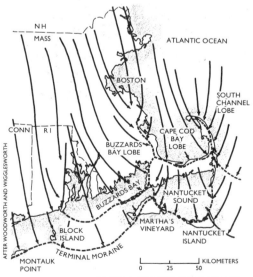

GLACIAL AND MODERN CAPE COD

NANTUCKET ISLAND, MASSACHUSETTS

Nantucket Island was formed from a glacial moraine earlier than that which formed Cape **Cod, and it too has been pounded and carved into its present shape by Atlantic waves. Until the War of 1812 against the** **British it was a haven for a hundred or more whaling ships at a time.**

bay afforded was offset by the difficulty of its shallow water. Until the *Mayflower*'s shallop had been rebuilt after stowage and use during the Atlantic crossing, the only way to get ashore was to wade a considerable distance: The November weather of 1620 was bitter and the icy sea unrelieved by the warmth of the Gulf Stream. The natives were unfriendly; for they had already seen too many Europeans and had good reason to be fearful of intruders. To compound the fears, the Separatists made the error of appropriating a cache of Indian corn: They were in the depths of low spirits but were "marvelously glad and their hearts encouraged" by their find. There was an unfortunate clash of arms after the event. (They later remedied this wrong and returned the corn in good measure.)

The cape was obviously unsuitable for settlement, but further exploration of what proved to be the mainland on the western shore of the almost circular Cape Cod Bay revealed land recently cleared yet devoid of people. The explanation was all too simple: The aboriginals had died wholesale from an epidemic of European diseases passed on by previous visitors. They had no immunity to European diseases, as the Europeans did not have immunity from water-borne and insect-borne North American diseases. Because of lack of time and facility to build shelter ashore, many *Mayflower* passengers wintered on the ship. During that first winter, as William Bradford recorded, "That which was most sad and lamentable was, that in two or three months' time half their company died, especially in January and February, being the depth of winter, and wanting houses and other comforts, being affected with scurvy and other diseases which this long voyage and their inaccommodate condition had brought upon them. So as there died some times two or three of a day in the forsaid time, that of one hundred and odd persons, scarce fifty remained." Like the Norse of Vinland and the Virginians of Roanoke and of the James River in Chesapeake Bay it took the New Englanders time and experience to adjust to a new continent. It took generations to become American.

Nantucket Island is 40 kilometers south of Cape Cod. Nauset Bay (pictured at right) is one of the regions of Cape Cod in which aboriginal people were concentrated. When the 17th-century explorer Samuel de Champlain visited the region in 1605, he found Nauset Bay to be well populated by aboriginals. This suggests that there existed a large number of stockaded dwellings at Nauset Bay and that the ground within the stockades was well cultivated.

Nauset Bay entrance is also thought to have been the site of the *Mayflower*'s landfall of Cape Cod in November, 1620, and the point at which the ship turned south. There are almost continuous shoals between Nauset Bay and Nantucket Island, and discovery of this fact may well have discouraged the *Mayflower* from continuing south. The decision was made to return north to what is now Provincetown Harbor. The Pilgrims got a hostile reception from the local Indian population on Cape Cod; the Indians had good cause to be wary and suspicious of new arrivals.

A month later, in December, 1620, having done some initial exploration, the Pilgrims sailed in the *Mayflower* across to the western shore of Cape Cod Bay and to the present site of Plymouth Massachusetts. The Pilgrims found cleared land there that had been inhabited by aboriginal people. The Indians had prepared the land but had died from European diseases, against which they had no immunity.

After a winter of terrible hardship, during which more than half their number died, the Pilgrims founded and built "Plimoth Plantation" (which has been reconstructed in modern times). In the autumn of 1621 William Bradford, then governor of Plymouth Colony, called for a day of "thanksgiving and prayer" after the Pilgrims' first harvest — a harvest made possible by the help that they had received from a lone but friendly Indian who had strayed into their plantation in March, 1621. They called the Indian Squanto; he is thought to have learned rudimentary English when he had been taken captive to England in 1605 and later returned to North America.

Plymouth Rock, also pictured here, later became a symbol of the Pilgrims' new-found liberty in the New World. Plymouth Rock had traveled far; for it is a glacial erratic. Like the Pilgrims themselves it had moved from one part of Avalonia to another.

NAUSET BAY, MASSACHUSETTS

THE MAYFLOWER, RECONSTRUCTED

PLYMOUTH PLANTATION, RECONSTRUCTED

PLYMOUTH ROCK

MONA SCHIST CLIFFS, ANGLESEY, NORTH WALES

As I sat shaded from the midmorning sunshine by the trees on the bank of Scrooby Manor's moat, I thought of Brewster, Bradford, Robinson, and their brethren. After their sojourn in Leiden, the Paris of the Netherlands at that time, they had determined to exchange the mullioned windows, the warm wide hearths, and the soft comfort that they had known so well at Scrooby Manor for stark, unknown regions of the New World. Incredible, I thought: Those people *truly* had had the courage of their convictions.

Some 150 kilometers to the southwest of Separatist Scrooby in central England there lay the fossilized remains of a small creature encased in rock and undisturbed for over 500 million years. Only in recent times has the true significance of this fossil and related fossil fauna in Britain and elsewhere around the North Atlantic been recognized by geologists. The fossil was of *Paradoxides*, a trilobite that matches others in the rocks of Avalonia, on which the future Pilgrim Fathers and later the Puritans of the Massachusetts Bay Company built their lives, their towns, their cities, their universities, and their industries. The Avalonian terrain of New England was cast in the same mold as the Avalonian terrain of southern Britain and Ireland. The only difference between these two now widely separated

AVALONIAN BASALT DIKE

These cliffs, photographed from South Stack near Holyhead on the coast of Anglesey, North Wales, are folded and metamorphosed sedimentary rocks called Mona Schists. They were formed about 625 million years ago and match similar Avalonian rocks that underlie southern Newfoundland and other regions of the East Coast of North America, now 3,000 kilometers or more distant and separated by the Mid-Atlantic Ridge and the Atlantic Ocean.

The smaller picture was also taken on an Anglesey seashore: It is of an Avalonian basalt formation. Nearby pillow lavas suggest that these rocks may once have formed part of the rifting Iapetus Ocean bed, the ocean that opened and closed before the present North Atlantic Ocean existed.

parts of Avalonia is that the rocks of New England traveled in advance of the Pilgrims. The Avalonian microcontinent from which both parts are formed was wrenched apart by rifting during the formation of the Iapetus Ocean, the proto–Atlantic Ocean. A large segment of coastal New England is part of the Avalonian microcontinent and, in common with the Avalon Peninsula of Newfoundland and other regions of the North American East Coast, is considered to be suspect terrain.

Avalonian rocks have been much studied in Britain. It was in fact Fred Dunning, the curator of the Geological Museum in London, who had first suggested to me that I should visit the Harlech Dome in North Wales to find *Paradoxides* in slaty Cambrian shales and manganese in underlying lagoon sediments. He had suggested, I remembered well, that I "might find the other half of the Harlech manganiferous lagoon in Newfoundland!" Dunning had also suggested that I should look closely at the rocks of Anglesey and particu-

THE ATLANTIC REALM OF PARADOXIDES

The now-scattered pieces of a certain style of suspect terrain distributed around North Atlantic shorelines once formed a single microcontinent: Avalonia. This microcontinent was an archipelago in the Iapetus Ocean, and in some of its shallow marine environments there existed the earliest known form of invertebrate life, the small, hard-shelled creatures called trilobites.

One species of Avalonian trilobite is called *Paradoxides*. After a geologically brief existence in Avalonian environments this species became extinct. But before extinction millions of such trilobites left their exoskeletons to fossilize in Avalonian sediments on the margin of a rifting and widening Iapetus Ocean.

During a long and complex geological history, first in the Iapetus Ocean cycle of events and then during the formation of the present North Atlantic Ocean, Avalonia was split up into fragments. Some of these fragments were large; some were

small. And some have been identified mixed in with other rocks scattered about the North Atlantic — from about as far northeast as the British Isles and as far southwest as the Carolina states.

In fact, wherever fragments of suspect terrain are found that contain the fossil *Paradoxides*, those terrains can have originated only in Avalonia. As shown in the illustration, adapted from a map kindly provided by Sean O'Brien

and his associates of the Canadian Survey, Newfoundland, such fragments of Avalonia underlie or can be seen on the surface of parts of England, North Wales, Southern Ireland, the Avalon Peninsula of Newfoundland, New Brunswick, the Boston Basin, and the East Coast piedmont. Collectively, the black areas once formed Avalonia — now "the Atlantic Realm of *Paradoxides*."

larly at the Precambrian rocks near Holyhead, the Mona Schists. He had told me that the latter are thought to be characteristic of the basement rocks that underlie some Avalonian rocks. I had since understood that both Avalonian rocks and formations like the Mona Schists are characterized by their "volcanic style" and that, as a consequence of the presence of such igneous intrusives, British geologists have been able to provide a time frame for the tectonic events that split up the Avalonian microcontinent 680 to 520 million years ago.

The Avalonian belt could equally be called "the Atlantic Realm of *Paradoxides*." The Avalonian terrains are believed to be the remnants of a late Precambrian and early pre-Pangean microcontinent. The New England fragments of this belt and similar fragments associated with North America did not reach their present positions until about 300 million years ago. Later, their relationship with cratonic rocks was further complicated by the continental collisions that resulted in the formation of Pangea. But between 680 and 520 million years ago, when all the Avalonian terrains were together, they would have appeared in the form of a strip of combined continental and oceanic fragments overlain by sedimentary rocks. This microcontinent was not just an extension of proto-Northern Europe but

probably an archipelago between Euro-Africa and proto-North America.

There were mountain-forming regions on both sides of the archipelago. On one side of the strip there were manganiferous lagoons and shallow slate-forming environments. These rocks are now to be seen in the Harlech Dome, in Southern Ireland, in the Avalon Peninsula of Newfoundland, in New Brunswick, in Maine, and in North Carolina. On the other side of the archipelago there were Precambrian sedimentary wedges of the kind to be seen today in Anglesey and in the Boston Basin of Massachusetts: These are called "Monian Schists." There is even one chunk of suspect terrain in the New World, at Newport, Rhode Island, that is considered by some geologists to be an excellent geological reproduction of Anglesey.

My legs were cramped from sitting too long. The wind rustling the trees was chill; it was time to leave Scrooby Manor's moat. I had learned at least one thing during the last year or two, I thought, as I ducked under the barbed wire fence: *Paradoxides* is a trilobite that will never be found in the rocks below the limestone blocks perched precipitously on the South Rim of the Grand Canyon of the Colorado — no matter how hard one contemplates the view.

ACKNOWLEDGMENTS AND BIBLIOGRAPHY

WORK on this book started in October, 1980, with a preliminary bout of photography and research that took me from Wyoming through Idaho and down to southern Utah and then north and east over the Rocky Mountains to Colorado — and so to England to complete a similar initial program. One objective at this stage was to talk over my ideas for the book with Frederick W. Dunning, the curator of the Geological Museum in London. He looked at me quizzically and, having told me something of the complex geology of the North American continent's Pacific Coast, wondered if perhaps I was about to bite off a little more than I could comfortably chew.

Of course he was right. Tectonic science is complex and difficult but quite the most exciting subject for a book that I had ever encountered. For example, there was the mind-boggling thought that most of the Pacific Coast is newly acquired terrain that did not originate as part of the North American continent but came from somewhere else in the Pacific Ocean. Perhaps unintentionally Fred Dunning fed the flames of enthusiasm with an outline of other and more ancient suspect terrain that related North Wales with Newfoundland and Norway with Greenland. I shall always be grateful to him for that marvelous introduction to a new world and to a new science.

North America is of considerable extent, so considerable as to be a daunting prospect for a singlehanded researcher and photographer. The key, I felt certain, would be my introduction to the professional world of geology in the United States and in Canada.

David A. Bentzin, a consulting petroleum geologist in Denver, Colorado, was enthused (I think) by the nature of my proposed project. He set up a meeting for me with the then president of the American Geological Institute, John D. Haun, and the two candidates for the next presidency of the American Association of Petroleum Geologists, Robert J. Weimer, of the Colorado School of Mines, and John M. Parker (who was in fact later elected). All four men spent an afternoon listening to my plans and to a summary of some of the hurdles to their fulfillment. The consequence of the meeting was support, introductions, and encouragement of a kind unparalleled in my experience. To John Haun in particular and to Bob Wymer and Jack Parker too I owe a great deal. But it is to Dave Bentzin that I owe the key introduction. How can one thank such generous people? I can only hope that this book justifies their time and their confidence.

The United States Geological Survey puts extraordinary effort and financial support into geological research. The result is an enormous fund of scientific papers and information about every aspect of geologic science in America. And more important perhaps is the availability of career scientists who have specialized in particular aspects of geology for most of their professional lifetimes. These scientists know remote regions of their country and other parts of the world like the backs of their hands and are ever willing to help even foreign laymen who have a serious interest in their work. In the course of researching this book I met and spent long periods of time with many such scientists, and their personal contribution has often been critical to my understanding of a broader aspect of their work. I am deeply in their debt and would like to express my thanks to them through Regional Geologists G. Brent Dalrymple of the USGS at Menlo Park, California, and Richard W. Mast of the USGS at Denver, Colorado. Individual acknowledgments are made before each chapter bibliography.

Canada is an enormous country with some of the most remote, interesting, and spectacular geological features on Earth. William H. Matthews III, Director of Education for the AGI, wrote on my behalf to W. W. Hutchison, then Director of the Geological Survey of Canada in Ottawa. As a result William G. Morgan of the CGS was asked to act as my guide, philosopher, and friend in the Canadian geological wilderness. Bill Morgan's patient but untiring efforts made possible meetings with leading specialists working in Newfoundland, the Northwest Territories, and the Northern Yukon, as well as in Ontario and Quebec. Again I am indebted to all these very generous people and to Bill Morgan and Bill Matthews in particular.

The archeology of ancient man in North America is a subject of profound interest to me and very much interlinked with the geology and glaciology of remote parts of the continent — except of course Meadowcroft Shelter near Pittsburgh, Pennsylvania. For archeological connections I am indebted first to He-len Pustmueller of the University of Denver and to Marie Wormington of Colorado College, Colorado Springs, who directed me wisely and well. I am particularly grateful for their introductions to other specialists in Canada and in the United States.

Certain chapters from the book are the subject of three one-hour BBC/PBS films to be broadcast at about the time of publication in Britain and in the United States. I first mentioned the problems of obtaining sponsorship for PBS co-production of these films in the United States to a good friend of mine, John C. Finlayson, of Rogart, Scotland, who was visiting America while I was working on the early stages of the book. Within a few months the first part of the dual sponsorship funding for the PBS co-production had been underwritten. I would like to pay a very special tribute to John Finlayson's support: Without it I doubt that the films would have been made. But this was not all: John later arranged for a light airplane from which I photographed the nether regions of Loch Ness and the Great Glen in Scotland. Obtaining such transport in North Scotland is a minor miracle in itself.

The vast region of the American Southwest that was to be the subject of both book and films had first to be surveyed. The alternatives were either to spend months on preliminary ground work or to do the survey from the air. Martin Litton of Portola, California, most generously made himself and his Cessna 175 available, first to do difficult and precise photographic work for the book and later for me to do a preliminary survey with BBC Producer Mike Andrews. No film-maker could have had a better introduction to the Southwest, and I could not have had a better and more generous friend and pilot than Martin Litton. He must have flown over 10,000 kilometers in these long sorties over some very difficult country.

In total several hundred people helped me in one important aspect or another of the research and the photography required for this book. I have here thanked those whose role was mainly introductory and advisory. But the fundamental details were supplied by a host of equally generous people. I regret that I can do no more here than to thank them most sincerely for their patience and their time. There is far more behind the further acknowledgments and bibliography that follow than is obvious from them.

My editor, Roger Jellinek, and my technical editor, Thomas L. T. Grose (Colorado School of Mines), have been enormously helpful. Jellinek has been in turn supportive or critical, agreeable or argumentative, but always available and always helpful during times of difficulty and tough deadlines. Tom Grose has gently pulled me back from more than one technical crevasse and has patiently guided me over several slippery slopes. Here too I was fortunate to have had an adviser who was always available and always willing to discuss a point. And there is Gary Hincks, whose extraordinary technical art decorates some of these pages: a quiet, delightful person and an easy man with whom to do very exacting work.

There is one final word of unstinted thanks that I gladly convey. It is to my wife, Joy, who has shared most of the pleasures of producing this book, many of the adventures, and all the difficulties. Her contribution has been of incalculable value.

General References

Bates, R. L., and Jackson, J. A., *Glossary of Geology*, American Geological Institute, Falls Church, VA, 1980.

Dott, Jr., R. H., et al., *Evolution of the Earth*, 3rd ed., McGraw-Hill, New York, 1981.

Eicher, D. L., et al., *History of the Earth*, Prentice-Hall, Englewood Cliffs, NJ, 1980.

The New Encyclopaedia Britannica, Encyclopaedia Britannica, Chicago, 1978.

Garner, H. F., *The Origin of Landscapes*, Oxford University Press, New York, 1974.

"Geological Highway Maps" (series of 11 maps), American Association of Petroleum Geologists, Tulsa, OK, 1979.

Hunt, C. B., *Natural Regions of the United States and Canada*, Freeman, San Francisco, 1974.

Mitchell, J. (ed.), *The Random House Encyclopedia*, Random House, New York, 1977.

Morison, S. E., *The European Discovery of America: The Northern Voyages*, Oxford University Press, New York, 1971.

Morison, S. E., *The European Discovery of America: The Southern Voyages*, Oxford University Press, New York, 1974.

Motz, L., *The Rediscovery of the Earth*, Van Nostrand Reinhold, New York, 1979.

Quinn, D. B., *America from Concept to Discovery: Early Exploration of North America*, Arno, New York, 1979.

Sanders, J. E., *Principles of Physical Geology*, Wiley, New York, 1981.

Smith, D. G. (ed.), *The Cambridge Encyclopedia of Earth Sciences*, Crown, New York, 1981.

Stearn, C. W., et al., *Geological Evolution of North America*, 3rd ed., Wiley, New York, 1979.

Trager, J. (ed.), *The People's Chronology*, Holt, Rinehart & Winston, New York, 1979.

CONTINENTS APART

Acknowledgments

Dr. Halldorsson, Armamagnean Institute, Reykjavik, Iceland, June, 1981.

Haukur Johannesson, Museum of Natural History, Reykjavik, Iceland, June, 1981.

Sean O'Brien and associates, Canadian Geological Survey, Nfld., July, 1981.

Earling Olafsson, Museum of Natural History, Reykjavik, Iceland, June, 1981.

Bjorn Ruriksson, Photographer, Reykjavik, Iceland, June, 1981.

James A. Tuck, Memorial University, St. John's, Nfld., July, 1981.

Knudsen Vilhjalmer, Film Producer, Reykjavik, Iceland, June, 1981.

Thorhallur Vilnundarson, Natural History Museum, Reykjavik, Iceland, June, 1981.

Dave Wilson, Geologist, Llangefni, Anglesey, Wales, June, 1981.

Bibliography

Anderson, J. G. C., and Owen, T. R., *The Structure of the British Isles*, 2nd ed., Pergamon, London, 1980.

East Pacific Rise Study Group, "Crustal Processes of the Mid-Ocean Ridge," *Science*, July, 1981.

Edwards, J. G. (ed.), *Snowdonia*, Her Majesty's Stationery Office, London, 1961.

Elms, E., "The Norse Site at L'Anse aux Meadows," Parks Canada, St. John's, Nfld., n.d.

Gass, I. G., "Ophiolites," *Scientific American*, August, 1982.

Ingstad, H., *Westward to Vinland*, St. Martin's, New York, 1965.

King, A. F., et al., "Geological and Mineralogical Associations of Canada" (Guidebook Field Trip B-6), Memorial University, St. John's, Nfld., May, 1974.

Millward, R., and Robinson, A., *Landscapes of North Wales*, David and Charles, Newton, Abbot, England, 1978.

Park Interpretation, "The Great Northern Peninsula," Parks Canada, St. John's, Nfld., n.d.

Saemundsson, K., et al., *Geology of Iceland*, National Energy Authority, Reykjavik, 1979.

Smith, B., and George, N. T., *British Regional Geology: North Wales*, Her Majesty's Stationery Office, London, 1961.

Tryggvason, E., "Rifting of the Plate Boundary in North Iceland, 1975-1978," National Energy Authority, Reykjavik, June, 1979.

Tuck, J. A., *Newfoundland and Labrador Prehistory*, Van Nostrand Reinhold, New York, 1976.

Williams, H., *Geological Development of the Northern Appalachians*, Memorial University, St. John's, Nfld., 1978.

THE NUCLEUS

Acknowledgments

Brenda Beebe, Archeologist, University of Toronto, Toronto, Ont., October, 1981.

Noel Broom, General Manager, Dome Petroleum Ltd., Calgary, Alta, August, 1981.

Cynthia C. Hill, Mayor, Inuvik, N.W.T., August, 1981.

P. F. Hoffman, Geological Survey of Canada, Ottawa, Ont., August, 1981.

J. Ross Mackay, Department of Geography, University of Columbia, Vancouver, B.C., August, 1981.

Wilfred Meyer, Geologist, Giant Mine, Yellowknife, N.W.T., August, 1981.

William C. Morgan, Geological Survey of Canada, Ottawa, Ont., 1981-1982.

Richard E. Morlan, Archeological Survey of Canada, Ottawa, Ont., October, 1982.

John D. Ostrick, Manager, Inuvik Research Laboratory, Dept. of Indian Affairs and Northern Development, N.W.T., August, 1982.

W. A. Padgham, Geologist, Canadian Geological Survey, Yellowknife, N.W.T., July, 1981.

Manley Showalter, Raecom Staking and Exploration Ltd., Yellowknife, N.W.T., July, 1981.

J. Tuzo Wilson, Director, Ontario Science Centre, Toronto, Ont., October, 1982.

Bibliography

Boyle, R. W., *The Geochemistry of Gold and Its Deposits*, Geological Survey of Canada, Hull, Que., 1980.

Hoffman, P. F., *Wopmay Orogen*, Geological Survey of Canada, Ottawa, 1980.

Irving, W. N., et al., "A Human Mandible: Pleistocene Faunal Assemblage: Eastern Beringia," *Canadian Journal of Archeology*, Toronto, 1977.

Jopling, A. V., Irving, W. N., and Beebe, B. F., "Stratigraphic, Sedimentological & Faunal Evidence for the Occurrence of pre-Sangamonian Artifacts in Northern Yukon," University of Toronto, Department of Geography, Toronto, March, 1981.

Mackay, J. R., "Pingos of the Tuktoyaktuk Peninsula Area, Northwest Territories," British Columbia Department of Geography, Vancouver, B.C., 1979.

Morlan, R.E., *Early Man in Northern Yukon Territory: Perspectives As of 1977*, Archeological Survey of Canada, Ottawa, 1978.

Morlan, R. E., *Pleistocene Archeology in Old Crow Basin: A Critical Reappraisal*, Archeological Survey of Canada, Ottawa, 1982.

Serson, P. H., *Tracking the North Magnetic Pole*, Department of Energy, Mines and Resources, Ottawa, 1980.

Stockwell, C. H., et al., *Geology of the Canadian Shield*, Department of Energy, Mines and Resources, Ottawa, 1970.

Wetherill, G. W., "The Formation of the Earth from Planetesimals," *Scientific American*, June, 1981.

UPPER CRUST

Acknowledgments

George Billingsley, U.S. Geological Survey, Flagstaff, AZ, 1981-1982.

Robert G. Breuning, Museum of Northern Arizona, Flagstaff, AZ, April, 1982.

Edwin H. Colbert, Paleontologist, Museum of Northern Arizona, Flagstaff, AZ, April, 1982.

Danny Davies, Dinosaur National Monument, Vernal, UT, April, 1982.

Thomas W. Gamble, Chief, Glen Canyon Field Division, U.S. Bureau of Reclamation, Page, AZ, 1980-1982.

Edwin D. McKee, U.S. Geological Survey, Denver, CO, 1978–

Bibliography

Breed, C. S., and Breed, W. J. (eds.), *Investigations in the Triassic Chinle Formation*, The Northern Arizona Society of Science & Art, Flagstaff, AZ, 1972.

Breed, W. J., *The Age of Dinosaurs in Northern Arizona*, The Northern Arizona Society of Science & Art, Flagstaff, AZ, 1975.

Colbert, E. H., *Dinosaurs*, Clarke, Irwin, Toronto, 1967.

Gunnerson, J. H., *The Fremont Culture*, The Peabody Museum, Cambridge, MA, 1969.

Lohman, S. W., "Arches National Park" (Bulletin 1393), U.S. Government Printing Office, Washington, 1975.

McKee, E. D., *Ancient Landscapes of the Grand Canyon Region*, Northland Press, Flagstaff, AZ, 1931.

McKee, E. D., et al., *Geology of the Grand Canyon*, 3rd ed., Museum of Northern Arizona, Flagstaff, AZ, 1978.

McKee, E. D., et al., "Fifth Field Conference, Powell Centennial River Expedition," Flagstaff, AZ, 1969 (reptd., 1972).

Sattler, H. R., *Dinosaurs of North America*, Lothrop, Lee & Shepard, New York, 1981.

Siegrist, R., *Prehistoric Petroglyphs and Pictographs in Utah*, Utah State Historical Society, Salt Lake City, UT, 1972.

Steel, R., and Harvey, A. P. (eds.), *The Encyclopedia of Prehistoric Life*, McGraw-Hill, New York, 1979.

CONTINENTAL DIVIDE

Acknowledgments

Warren Hamilton, U.S. Geological Survey, Denver, CO, 1981-1982.

Louise Kiteley, Consulting Geologist, Denver, CO, 1982.

Lorraine L. Mintzmyer, Regional Director, Rocky Mountain Region, National Park Service, Denver, CO, 1982.

Bibliography

Abbott, C., et al., *Colorado*, Colorado Associated University Press, Boulder, CO, 1982.

Barth, G., *Instant Cities*, Oxford University Press, New York, 1975.

Chronic, J., and Chronic, H., "Prairie Peak and Plateau" (Bulletin 32), Colorado Geological Survey, Denver, CO, 1972.

Dorsett, L. W., *The Queen City*, Pruett, Boulder, CO, 1977.

Hamilton, W., "Plate-tectonic Mechanism of Laramide Deformation," *Contributions to Geology,* University of Wyoming, January, 1981.

Holliday, J. S., *The World Rushed In*, Simon & Schuster, New York, 1981.

USGS Publications, "Denver's Geologic Setting," Denver, CO, 1980.

Vickers, W. B., *History of the City of Denver*, Baskin, Chicago, 1880.

SUSPECT TERRAIN

Acknowledgments

Frank E. Baker, Sohio Alaska Petroleum Company, Anchorage, AK, September, 1981.

Alec Brogle, Marine Biologist, Yakutat, AK, September, 1981.

J. G. Graham, Whitby Museum, Whitby, Yorks., June, 1982.

Bob Hartzler, Sohio Alaska Petroleum Company, Anchorage, AK, September, 1981.

Roger C. Herrara, Sohio Alaska Petroleum Company, Anchorage, AK, September, 1981.

Arnie J. Israelson, Chevron Agent, Yakutat, AK, September, 1981.

Mike Ivers, Gulf Air Taxi, Yakutat, AK, September, 1981.

Marvin D. Mangus, Consulting Geologist, Anchorage, AK, September, 1981.

Bob M. Mayer, Sohio Alaska Petroleum Company, Anchorage, AK, September, 1981.

Gil Mull, Alaska Geological Survey, Anchorage, AK, September, 1981.

Charles B. Towill, Sohio Alaska Petroleum Company, Anchorage, AK, September, 1981.

Bibliography

"Abstracts of the Fifth Biennial Meeting," University of Alberta, Edmonton, Alta., September, 1978.

"Arctic Steppe-Tundra: A Yukon Perspective," *Science*, June, 1980.

Bird, K. J., *Petroleum of the North Slope in Alaska*, U.S. Geological Survey, Menlo Park, CA, 1981.

Hays, W. W., *Facing Geologic and Hydrologic Hazards*, U.S. Government Printing Office, Washington, 1981.

Hopkins, D. M., *The Bering Land Bridge*, Stanford University Press, Stanford, CA, 1967.

Irving, W. N., and Harington, C. R., "Upper Pleistocene Radiocarbon-dated Artifacts from the Northern Yukon," *Science*, January, 1973.

Jones, D. L., et al., "The Growth of Western North America," *Scientific American*, November, 1980.

Laughlin, W. S. (ed.), *The First Americans*, Gustav-Fischer, New York, n.d.

"The Origin of the Arctic Ocean" (mini-symposium), The Alaska Geological Society, Anchorage, AK, April, 1981.

Pewe, T. L., *Quaternary Geology of Alaska*, U.S. Government Printing Office, Washington, 1975.

Phillips, J. W., "Prudhoe Bay and Duke Island," courtesy of G. Mull, n.d.

Roeder, D., and Mull, C. G., "Tectonics of Brooks Range Ophiolites, Alaska," American Association of Petroleum Geologists Bulletin, Tulsa, OK, September, 1978.

Shalkop, A. (ed.), "Exploration of Alaska" (Captain Cook Commemorative Lectures), Cook Inlet Historical Society, Anchorage, AK, 1980.

ACTIVE MARGIN

Acknowledgments

John T. Alfors, California Division of Mines and Geology, Sacramento, CA, April, 1982.

John D. Cherry, National Park Service, San Francisco, CA, April, 1982.

G. Brent Dalrymple, U.S. Geological Survey, Menlo Park, CA, April, 1982.

Dave Howell and Dave Jones, U.S. Geological Survey, Menlo Park, CA, April-October, 1982.

Glen R. McGowan, Department of Parks and Recreation, Monterey Area, CA, April, 1982.

Robert Mondavi, Napa Valley Winery, Napa Valley, CA, April, 1982.

Larry Quist, National Park Service, San Francisco, CA, April, 1982.

Clyde Ray, Department of Parks and Recreation, Monterey Area, CA, April, 1982.

Bibliography

Coney, P. J., Jones, D. L., and Monger, J. W. H., "Cordilleran Suspect Terranes," *Nature*, November, 1980.

Hamilton, W., "Mesozoic Tectonics of the Western United States," U.S. Geological Survey, Denver, CO, 1978.

Hamilton, W., *Plate Tectonics and Man* (USGS Annual Report), U.S. Geological Survey, Washington, 1976.

Jones, D. L., "Wrangellia—a Displaced Terrane in Northwestern North America," *Canadian Journal of Earth Sciences*, 1977.

Jones, D. L., et al., "Recognition, Character, and Analysis of Tectonostratigraphic Terranes in Western North America," U.S. Geological Survey, Menlo Park, CA, January, 1982.

Margolin, M., *The Ohlone Way*, Heyday, Berkeley, CA, 1968.

Miller, D. J., "Giant Waves in Lituya Bay, Alaska" U.S. Government Printing Office, Washington, 1960.

Norris, R. M., and Webb, R. W., *Geology of California*, Wiley, New York, 1976.

Quinn, D. B., *America from Concept to Discovery*, Arno, New York, 1979.

MOBILE BELT

Acknowledgments

Dan Davies, Arizona-Sonora Desert Museum, Tucson, AZ, April, 1982

Rich Delaney, Arizona-Sonora Desert Museum, Tucson, AZ, April, 1982.

Robert O. Fournier, U.S. Geological Survey, Menlo Park, CA, April, 1982.

Christopher Helms, Public Affairs Officer, Arizona-Sonora Desert Museum, Tucson, AZ, April, 1982.

Harold E. Malde, U.S. Geological Survey, Denver, CO, 1981-1982.

Richard S. Rayner, Death Valley National Monument, National Park Service, Furnace Creek, Death Valley, CA, April, 1982.

Bibliography

Bierley, C. L., "Microbiological Mining," *Scientific American*, August, 1982.

Brock, T. D., and Brock, L. M., *Life in the Geyser Basins*, Yellowstone Library & Museum Association, Madison, WI, 1978.

Christiansen, R. L., and McKee, E. H., "Late Cenozoic Volcanic and Tectonic Evolution of the Great Basin" (Paper 152), Geological Society of America, 1976.

Folsome, C. E., *The Origin of Life*, Freeman, San Francisco, 1979.

Helms, C. L., *The Sonoran Desert*, KC Publications, Las Vegas, NV, 1980.

Hildreth, W., *Death Valley Geology,* Death Valley Natural History Association, National Park Service, Death Valley, CA, 1976.

Hunt, C. B., *Death Valley*, University of California Press, Berkeley, CA, 1975.

Malde, H. E., "The Catastrophic Late Pleistocene Bonneville Floor in the Snake River Plain, Idaho," U.S. Government Printing Office, Washington, 1968.

Marler, G. D., *Studies of Geysers and Hot Springs Along the Firehole River*, Yellowstone Library & Museum Association, Madison, WI, 1978.

Newman, E. A., and Hartline, P. H., "The Infrared Vision of Snakes," *Scientific American*, March, 1982.

Schullery, P., *The Bears of Yellowstone*, Yellowstone Library & Museum Association, Yellowstone National Park, WY, 1980.

Snyder, E., *Arizona Field Guide*, Arizona State University, Lamont, AZ, 1978.

STABLE PLATFORM

Acknowledgments

Dick Cavagnaro, General Motors, Lockport, NY, October, 1982.

Ken Dodd, British Waterways, Bingley, Yorks., May, 1982.

Robert H. Dott, University of Wisconsin, Madison, WI, June, 1982.

James. C. Knox, University of Wisconsin, Madison, WI, June, 1982.

Jack Krajewski, New York State Geological Survey, Buffalo, NY, October, 1982.

John E. Kutzbach, Meteorologist, University of Wisconsin, Madison, WI, June, 1982.

David Lonsdale, John G. Shedd Aquarium, Chicago, IL, June, 1982.

William T. Moorman, Bureau of Parks and Recreation, Madison, WI, June, 1982.

W. H. Shank, American Canal and Transportation Center, York, PA, 1981-1982.

Don Sutherland, University of Edinburgh, Edinburgh, Scotland, May, 1982.

Bibliography

Dott, R. H., and Batten, R. L., *Evolution of the Earth*, McGraw-Hill, New York, 1981.

Fischer, W. A., *Earthquake*, Yellowstone Library & Museum Association, Yellowstone National Park, WY, 1976.

Flint, R. F., *Glacial and Quaternary Geology*, Wiley, New York, 1973.

Flint, R. F., *Glacial and Pleistocene Geology*, Wiley, New York, 1969.

Garner, H. F., *The Origin of Landscapes*, Oxford University Press, New York, 1974.

Imbrie, J., and Imbrie, K. P., *Ice Ages*, Enslow, Hillsdale, NJ, 1979.

Johnson, A. C., "A Major Earthquake Zone on the Mississippi," *Scientific American*, April, 1982.

Keefer, W. R., *The Geologic Story of Yellowstone National Park*, Yellowstone Library & Museum Association, Yellowstone National Park, WY, 1976.

"Lockport—to the Canal and Beyond," Lockport Area Chamber of Commerce, Lockport, NY, n.d.

Matsch, C. L., *North America and the Great Ice Age*, McGraw-Hill, New York, 1976.

Mayer, H. M., and Wade, R. C., *Chicago*, University of Chicago Press, Chicago, 1969.

Neale, J., and Flenley, J. (eds.), *The Quaternary in Britain*, Pergamon, Elmsford, NY, 1981.

Ostrom, M. E., "Quaternary History of the Driftless Area," Geological and Natural History Survey, University of Wisconsin Extension, Madison, WI, 1982.

Shank, W. H., *Towpaths to Tugboats*, The American Canal and Transportation Center, York, PA, 1982.

Sissons, J. B., "Late Glacial Marine Erosion and a Jökullhaup Deposit in the Beauly Firth," *Geology*, vol. 17, 1981.

Stearn, C. W., Carroll, R. L., and Clark, T. H., *Geological Evolution of North America*, Wiley, New York, 1979.

Tesmer, I. H. (ed.), *Colossal Cataract: The Geologic History of Niagara Falls*, State University of New York Press, Albany, NY, 1981.

Thorarinsson, S., and Seamundsson, K. (eds.), *Geology of Iceland*, Geoscience Society of Iceland, Reykjavik, l980.

FLUVIAL PLAIN

Acknowledgments

Frederic M. Chatry, U.S. Army Corps of Engineers, New Orleans, LA, June, 1982.

Clark C. Hawley, New Orleans Steamboat Co., New Orleans, LA, June, 1982.

Bill Iseminger, Cahokia Mounds Museum, East St. Louis, MO, June, 1982.

Charles Kolb, Consulting Geologist, Vicksburg, MS, June, l982.

Brian J. Mitchell, St. Louis University, St. Louis, MO, June, 1982.

Bruce A. Sossaman, Public Information Officer, U.S. Army Corps of Engineers, New Orleans, LA, June, 1982.

Samuel Wilson, Architect, New Orleans, LA, June, 1982.

Bibliography

Brain, J. P., "On the Tunica Trail" (Anthropological Study 1), Department of Culture, Recreation and Tourism, Baton Rouge, LA, June, 1977.

Christian, M., *Negro Ironworkers of Louisiana 1718–1900*, Pelican, Gretna, LA, 1972.

Clay, F. M., *A Century on the Mississippi*, U.S. Government Printing Office, Washington, 1976.

Cowdrey, A. E., *Land's End*, U.S. Government Printing Office, Washington, 1977.

Dobney, F. J., *River Engineers on the Middle Mississippi*, U.S. Government Printing Office, Washington, 1978.

Fisk, H.N., "Results of Geological Investigations of the Alluvial Valley of the Lower Mississippi River, December, 1945, to January, 1946," U.S. Government Printing Office, Washington, 1946.

Haines, A. L., *The Oregon Trail*, Patrice, MO, 1981.

Holmes, J. D. L., *The Story of Spanish Moss and the Wax Tree*, Hope Publications, Jefferson, LA, n.d.

Iseminger, W., *Ancient Illinois: Cahokia Mounds*, Cahokia Mounds Museum Society, East St. Louis, MO, 1974-1980.

Johnston, A. C., "A Major Earthquake Zone on the Mississippi," *Scientific American*, April, 1982.

Kane, M. F., Hildenbrand, T. G., and Hendricks, J. D., "A Model for the Tectonic Evolution of the Mississippi Embayment," U.S. Geological Survey, Denver, CO, 1981.

Kolb, C. R., "Geological Notes on Vicksburg and the Yazoo Basin" (Paper A), AAPG/SEPM Annual Convention, May, 1976.

Kolb, C. R., and Van Lopik, J. R., "Depositional Environments of the Mississippi River Deltaic Plain," Houston Geological Society, Houston, TX, 1976.

Kolb, C. R., "Paper presented at the Gulf Coast Association of Geological Societies," Lafayette, LA, October, 1980.

Mayer, H. M., and Wade, R. C., *Chicago*, University of Chicago Press, Chicago, 1969.

Russ, D. P., "Late Holocene Faulting and Earthquake Recurrence in the Reelfoot Lake Area," U. S. Geological Survey, Denver, CO, December, 1979.

Stegner, W., *Beyond the Hundredth Meridian*, Houghton Mifflin, Boston, 1954.

Taylor, J. G., *Louisiana*, Norton, New York, 1976.

Wilson, Jr., S., *The Vieux Carré, New Orleans*, City of New Orleans, New Orleans, LA, l968.

Wilson, Jr., S., and Huber, L. V., *The Cabildo*, Pelican, Gretna, LA, 1973.

Zoback, M. D., et al., "Recurrent Intraplate Tectonism in the New Madrid Seismic Zone," *Science*, August, 1980.

PASSIVE MARGIN

Acknowledgments

Paul Albany, Historian, Nassau, Bahamas, May, 1982.

Rodney Attill, Bahamas National Trust, Nassau, Bahamas, May, 1982.

R. W. Boebel, Consulting Petroleum Geologist, New Orleans, LA, June, 1982.

Richard V. Cant, Geologist, Ministry of Works, Nassau, Bahamas, May, 1982.

Julian Granberry, Consultant Archeologist, Horseshoe Beach, FL, May, 1982.

Lee Hughes, *New York Times*, Miami, FL, May-June, 1982.

Ivor Noël Hume, Archeologist, Colonial Williamsburg, Williamsburg, VA, June, 1982.

Gerald Johnson, College of William & Mary, Williamsburg, VA, June, 1982.

Mark Questell, Watchtower Society, Brooklyn, NY, July, 1982.

Oris S. Russell, Ministry of External Affairs, Nassau, Bahamas, May, 1982.

John V. Sanders, Postmaster General, Nassau, Bahamas, May, 1982.

Art Smith, AV Director, Colonial Williamsburg, Williamsburg, VA, June, 1982.

Mark Wenger, Architect, Colonial Williamsburg, Williamsburg, VA, June, 1982.

Bibliography

Adams, R. W., et al., *Field Guide Geology of San Salvador*, CCFL Bahamian Field Station, Miami, FL, 1981.

Campbell, D. G., *The Ephemeral Islands*, Macmillan, London, 1978.

Cummings, A. L., *Architecture in Early New England*, Old Sturbrige, MA, 1976.

Current, R. N., Williams, T. H., and Freidel, F., *American History*, Knopf, New York, 1979.

Dietz, R. S., et al., *Geotectonic Evolution and Subsidence of Bahama Platform,* MOAA Laboratories, Miami, FL, 1972.

Gascoyne, M., et al., "Sea-level Lowering during the Illinoian Glaciation," *Science*, 1979.

Gebelein, C. D., *Guidebook for Modern Bahamian Platform Environments*, 2nd ed., University of California Press, Berkeley, CA, 1974.

Hoffmeister, J. E., *Land from the Sea*, University of Miami Press, Miami, FL, 1974.

Hume, I. N., *Martin's Hundred*, Knopf, New York, 1982.

Johnson, H., *Trees*, Beazley, London, 1977.

"Legacy from the Past," Colonial Williamsburg Foundation, Williamsburg, VA, 1978.

Meyerhoff, A. A. (ed.), *Geology of Natural Gas in South Louisiana*, Lafayette and New Orleans Geological Society, Baton Rouge and New Orleans, LA, 1966.

Rouse, I., "Prehistory of the West Indies," *Science,* May, 1964.

Uchupi, E., et al., *Structure and Origin of Southeastern Bahamas*, American Association of Petroleum Geologists, Woods Hole, MA, 1971.

APPALACHIA

Acknowledgments

James M. Adovasio, University of Pittsburgh, Pittsburgh, PA, June, 1982.

Thomas H. Anderson, University of Pittsburgh, Pittsburgh, PA, June, 1982.

Louise Ferguson, Pittsburgh History & Landmarks Foundation, Pittsburgh, PA, June, 1982.

Norman K. Flint, University of Pittsburgh, Pittsburgh, PA, June, 1982.

John T. Galey, Consulting Petroleum Geologist, Somerset, PA, June, 1982.

Samuel P. Hays, University of Pittsburgh, Pittsburgh, PA, June, 1982.

John A. Harper, Pittsburgh Topographic and Geologic Survey, Pittsburgh, PA, June, 1982.

Herbert Johl, Ellicott City B & O Railroad Museum, Colombia, MD, June, 1982.

Walter C. Kidney, Journalist, Pittsburgh, PA, June, 1982.

A. R. D. Perrins, Reynolds Metals Company, Richmond, VA, June, 1982.

Harold B. Rollins, University of Pittsburgh, Pittsburgh, PA, June, 1982.

John E. Sanders, Columbia University, New York, NY, July, 1982.

Michael Siegal, University of Pittsburgh, Pittsburgh, PA, June, 1982.

Marion Smith, Railroad Museum, Baltimore, MD, June, 1982.

Bibliography

Bally, A. W., et al., *Geology of Passive Continental Margins*, American Association of Petroleum Geologists, Tulsa, OK, 1981.

Carlisle, R. C., and Adovasio, J. M. (eds.), *Meadowcroft Collected Papers*, Society for American Archeology, Minneapolis, MN, 1982.

Cook, F. A., et al., "The Southern Appalachians and the Growth of Continents," *Scientific American*, October, 1980.

DeGolyer, E., *The Journey of Three Englishmen across Texas*, Peripatetic, El Paso, TX, 1947.

Harris, L. D., et al., "Evaluation of Southern Eastern Overthrust Belt beneath Blue Ridge-Piedmont Thrust," American Association of Petroleum Geologists, Tulsa, OK, 1981.

Haun, J. D., "Coal" (unpublished papers), Denver, CO, 1983.

Johl, H. H., "Historic Structures Report: The First Terminus of the Baltimore & Ohio Railroad," The Ellicott City B & O Railroad Station Museum, Ellicott, MD, 1982.

Kirkwood, J. J., *Waterway to the West*, Eastern National Park and Monuments Association, Blue Ridge Parkway, VA, 1963.

Murphy, P., *Richmond and Its Historic Canals*, Reynolds Metals, Richmond, VA, 1971.

Pitman, III, W. C., and Golovchenko, X., "The Role of Sea Level Change in the Shaping of the Appalachian Mountains," American Association of Petroleum Geologists, New York, 1981.

Ransom, P. J. G., *The Archaeology of Railways*, Windmill Press, Kingswood, Surrey, 1981.

St. George, J., *The Brooklyn Bridge*, Putnam's, New York, 1982.

Salvadori, M., *Why Buildings Stand Up*, Norton, New York, 1980.

Schlee, J. S., "A Comparison of Two Atlantic-type Continental Margins," U.S. Government Printing Office, Washington, 1980.

Shank, W. H., *Three Hundred Years with the Pennsylvania Traveler*, American Canal & Transport Center, York, PA, 1982.

Watts, A. B., *The U.S. Atlantic Continental Margin*, American Association of Petroleum Geologists, New York, 1981.

Williams, H., "Geological Development of the Northern Appalachians," St. John's, Nfld., 1980.

Wilshusen, J. P., *Geologic Hazards in Pennsylvania*, Pennsylvania Geological Survey, Harrisburg, PA, 1979.

Windley, B. F., *The Evolving Continents*, Wiley, New York, 1977.

AVALONIA

Acknowledgments

James R. Allen, National Park Service, Boston, MA, October, 1982.

James W. Baker, Plimoth Plantation, Plymouth, MA, October, 1982.

John Debo and Lawrence D. Gall, Lowell National Historic Park, Lowell, MA, October, 1982.

Malcolm J. Dolby, Doncaster Museum, Doncaster, Yorks., May, 1982.

Jim G. Gott, Saugus Iron Works National Historic Site, Saugus, MA, October, 1982.

Malcolm Hill, Northeastern University, Boston, MA, October, 1982.

Steven H. Lewis, Deputy Regional Director, National Park Service, Boston, MA, October, 1982.

Francis P. McManamon, Archeologist, National Park Service, Boston, MA, October, 1982.

James W. Skehan, Geologist, Weston Observatory, Boston, MA, October, 1982.

Bibliography

Dexter, H. M. (ed.), *Mourt's Relation*, Garrett Press, 1969 (rept.).

Dolby, M. J., "Scrooby Information Sheet," Scrooby Parish Council, Scrooby, Lincolns., 1981.

Dunning, F. W., et al., *Britain before Man*, Her Majesty's Stationery Office, London, 1978.

Huestis, R. T., *The Pilgrims: Book One, Group Portrait*, Cove House, Pigeon Cove, MA, 1966.

Jeremy, D. J., *Transatlantic Industrial Revolution*, MIT Press, Cambridge, MA, 1981.

Kent, D. V., "Paleomagnetic Evidence for Post-Devonian Displacement of the Avalon Platform (Newfoundland)," Lamont-Doherty Geological Observatory, Palisades, NY, 1981.

Lewis, W. D., *Iron and Steel in America*, Hagley Museum, Hagley, MA, 1981.

McManamon, F. P., "Prehistoric Land Use on Outer Cape Cod," National Park Service, Boston, MA, 1980.

McManamon, F. P., "The Archeology of Cape Cod National Seashore," National Park Service, Boston, MA, 1982.

Matthews, L. H., *The Natural History of the Whale*, Columbia University Press, New York, 1978.

Rast, N., *The Avalonian Plate in the Northern Appalachians and Caledonides*, University of Kentucky Press, Lexington, KY, 1981.

Rast, N., and Skehan, J. W., "Possible Correlation of Precambrian Rocks of Newport, Rhode Island with Those of Anglesey, Wales," *Geology*, December, 1981.

Skehan, J. W., "Subduction Zone between the Paleo-American and the Paleo-African Plates in New England," *Geofisica Internacional*, vol. 13, no. 4, pp. 291–308, Mexico, D.F., 1973.

Skehan, J. W., and Murray, D. P., *Geologic Profile across Southeastern New England*, Elsevier, Amsterdam, February, 1980.

Strahler, A. N., *A Geologist's View of Cape Cod*, Natural History Press, New York, 1966.

Walton, I., *The Life of Mr. George Herbert*, Oxford University Press, London, 1675 (reptd., 1973).

INDEX